Springer Theses

Recognizing Outstanding Ph.D. Research

Aims and Scope

The series "Springer Theses" brings together a selection of the very best Ph.D. theses from around the world and across the physical sciences. Nominated and endorsed by two recognized specialists, each published volume has been selected for its scientific excellence and the high impact of its contents for the pertinent field of research. For greater accessibility to non-specialists, the published versions include an extended introduction, as well as a foreword by the student's supervisor explaining the special relevance of the work for the field. As a whole, the series will provide a valuable resource both for newcomers to the research fields described, and for other scientists seeking detailed background information on special questions. Finally, it provides an accredited documentation of the valuable contributions made by today's younger generation of scientists.

Theses are accepted into the series by invited nomination only and must fulfill all of the following criteria

- They must be written in good English.
- The topic should fall within the confines of Chemistry, Physics, Earth Sciences, Engineering and related interdisciplinary fields such as Materials, Nanoscience, Chemical Engineering, Complex Systems and Biophysics.
- The work reported in the thesis must represent a significant scientific advance.
- If the thesis includes previously published material, permission to reproduce this must be gained from the respective copyright holder.
- They must have been examined and passed during the 12 months prior to nomination.
- Each thesis should include a foreword by the supervisor outlining the significance of its content.
- The theses should have a clearly defined structure including an introduction accessible to scientists not expert in that particular field.

More information about this series at http://www.springer.com/series/8790

Raphael Hellwig

Alkyne-Based Nanostructures on Silver Substrates

Doctoral Thesis accepted by
the TU München, Garching, Germany

Springer

Author
Dr. Raphael Hellwig
Physics Department
TU München
Garching, Bayern, Germany

Supervisor
Prof. Dr. Johannes Barth
Physics Department
TU München
Garching, Bayern, Germany

ISSN 2190-5053 ISSN 2190-5061 (electronic)
Springer Theses
ISBN 978-3-030-00996-0 ISBN 978-3-030-00997-7 (eBook)
https://doi.org/10.1007/978-3-030-00997-7

Library of Congress Control Number: 2018955597

This Springer imprint is published by the registered company Springer Nature Switzerland AG
The registered company address is: Gewerbestrasse 11, 6330 Cham, Switzerland

Supervisor's Foreword

The exploration of molecular nanostructures and elementary reactions on solid surfaces underpins the development of novel materials and processes. These are relevant for numerous application fields including nanoelectronics, organic electronics, heterogenous catalysis, and sensing. By using well-defined substrates in an ultra-high vacuum environment, the investigated systems can be characterized down to the molecular level, while at the same time innovative chemical transformations and nanofabrication schemes are unveiled. Frequently, a combination of local investigations at the molecular scale by scanning probe microscopy with space-averaging photoemission spectroscopy tools is employed, whereby computational modeling with density functional theory can be very helpful to guide the data interpretation. Accordingly, the present work was dedicated to exploring a very promising class of compounds and reaction pathways in this regard, namely toward realizing novel carbon scaffolds or organometallic nanostructures. Specifically, the pioneering investigations on homocoupling of alkynes reveal a promising approach toward the fabrication of graphyne or graphdiyne networks. These are expected to provide remarkable electronic properties and thus extend the application spectrum of the related graphene layers. Beyond targeting novel two-dimensional materials, it was also demonstrated that a suitable choice of organic precursors and reaction protocols can yield one-dimensional graphdiyne nanowires. Moreover, the interaction of single adatoms with alkyne units gives rise to intriguing metal-acetylide formations or even mediates covalent bond formations if rare-earth species such as holmium are employed.

With his dissertation, Raphael Hellwig beautifully illustrates a powerful and versatile multitechnique approach toward advancing molecular chemistry at interfaces and realizing novel functional materials in reduced dimensions. The thesis not only comprises a series of original scientific results and fascinating insights, but also features a concise and readable introduction to the field and the underlying experimental methodology.

Munich, Germany
July 2018

Prof. Dr. Johannes Barth

Abstract

Acetylene-based precursors are essential for organic synthesis protocols and represent promising building blocks for the creation of low-dimensional carbon materials different from graphene. While their reactivity toward covalent carbon–carbon bond formation is well studied in solution, only limited efforts have been undertaken to create alkyne-based nanostructures on metal surfaces under ultra-high vacuum conditions. Within this thesis, thoughtfully designed alkyne monomers are utilized for the bottom-up creation and scanning tunneling microscopy (STM) characterization of on-surface formed organometallic and covalent nanostructures on single-crystal silver substrates.

Cyano-functionalized organometallic Ag-bis-acetylide nanowires are created through the heat-stimulated coupling of a terminal alkyne precursor equipped with polar side groups (CN-DETP). Complementary to STM, X-ray spectroscopy delivers space-averaged information on their conformational and organometallic bonding aspects. While Ag(110) mediates the formation of dense-packed domains of metallopolymer molecular wires arranged via supramolecular side group recognition, the isotropic Ag(111) surface steers the formation of extended graphdiyne nanowires. The highly selective on-surface polymerization mechanism on Ag(111) emanates from the polar side groups of CN-DETP, which amplify the chemoselectivity toward terminal alkyne homocoupling. STM manipulations on isolated nanochains reveal extraordinary mechanical flexibility and isomer switching characteristics with prospects for high-density information storage. Besides the on-surface synthesis of linear polymer chains, dinuclear cobalt alkynyl complexes are created by employing a hexaethynylbenzene derivative co-deposited with cobalt (Co) atoms on Ag(111). STM results supported by density functional theory modeling reveal the expression of an Archimedean tiling, which is attributed to the epitaxy-induced adsorption of organocobalt complexes. Moreover, single-molecule manipulation experiments on dinuclear compounds reveal enantiomeric switching between their chiral conformer states and Kondo screening after STM tip-based activation. Replacing the 3D transition element Co by the 4f-block element holmium (Ho), we demonstrate that selective metal alkynyl bonding is driving the formation of surface-supported organolanthanide complexes. Furthermore, we report on the ability of surface-supported Ho atoms to catalyze terminal alkyne dimerization and trimerization on Ag(111); i.e, open porous polymer structures are created via a Ho-catalyzed cyclotrimerization reaction proceeding at 200 K. The results of this thesis on the characterization of alkyne-based on-surface nanostructures open up new perspectives in the fields of heterogenous catalysis and synthesis of functional materials.

List of Publications

R. Hellwig, T. Paintner, Z. Chen, M. Ruben, A. Seitsonen, F. Klappenberger, H. Brune and J. V. Barth. "Epitaxy-induced assembly and enantiomeric switching of an on-surface formed dinuclear organocobalt complex" *ACS Nano,* vol. 11, pp. 1347-1359, 2017.

R. Hellwig, M. Uphoff, T. Paintner, J. Björk, M. Ruben, F. Klappenberger and J. V. Barth. "Ho-mediated alkyne reactions at low temperatures on Ag(111)" *Chem Eur. J.,* 10.1002/chem.201803102, 2018.

R. Hellwig, M. Uphoff, Y.-Q. Zhang, P. Du, M. Ruben, F. Klappenberger and J. V. Barth. "Cyano-functionalized Ag-bis-acetylide nanowires on Ag(110)" *to be published.*

P. B. Weber, R. Hellwig, T. Paintner, M. Lattelais, M. Paszkiewicz, P. C. Aguilar, P. S. Deimel, Y. Guo, Y.-Q. Zhang, F. Allegretti, A. C. Papageorgiou, J. Reichert, S. Klyatskaya, M. Ruben, J. V. Barth, M.-L. Bouquet and F. Klappenberger. "Surface-guided formation of an organocobalt complex" *Angew. Chem. Int. Ed.*, vol. 55, pp. 5754-5759, 2016.

F. Klappenberger, R. Hellwig, P. Du, T. Paintner, M. Uphoff, L. Zhang, T. Lin, B. A. Moghanaki, M. Paszkiewicz, I. Vobornik, J. Fujii, O. Fuhr, Y.-Q. Zhang, F. Allegretti, M. Ruben and J. V. Barth. "Functionalized graphdiyne nanowires: on-surface synthesis and assessment of band structure, flexibility, and information storage potential" *Small*, vol. 14, pp. 1704321 (1-10), 2018.

Y.-Q. Zhang, T. Paintner, R. Hellwig, S. Klyatskaya, M. Ruben, J. V. Barth, F. Klappenberger, "Interfacial synthesis of micrometer-scale regular silver-intercalated graphdiyne network assisted by heterogeneous catalysis", *in preparation.*

Y.-Q. Zhang, T. Lin, B. Cirera, R. Hellwig, C.-A. Palma, Z. Chen, M. Ruben and J. V. Barth. "One-dimensionally disordered chiral sorting via racemic tiling in surface-confined supramolecular assembly of achiral tectons" *Angew. Chem. Int Ed.*, vol. 56, pp. 7797-7802, 2017.

M. Uphoff, R. Hellwig, G. S. Michelitsch, K. Reuter, F. Klappenberger, H. Brune and J. V. Barth. "Assembly of robust Holmium-directed 2D metal-organic coordination complexes and networks on the Ag(100) surface" *submitted.*

Y.-Q. Zhang, J. Björk, P. Weber, R. Hellwig, K. Diller, A. C. Papageorgiou, S. Cheol Oh, S. Fischer, F. Allegretti, S. Klyatskaya, M. Ruben, J. V. Barth and F. Klappenberger. "Unusual deprotonated alkynyl hydrogen bonding in metal-supported hydrocarbon assembly" *J. Phys. Chem. C,* 2015, vol. 119 (17), pp. 9669-9679, 2015.

K. Diller, A. Singha, M. Pivetta, S. Rusponi, C. Wäckerlin, R. Hellwig, A. Verdini, A. Cossaro, L. Floreano, E. Vélez-Fort, J. Dreiser and H. Brune. "On-surface synthesis of single-molecule magnets" *in preparation.*

T. Paintner, J. Björk, P. Du, S. Klyatskaya, M. Paszkiewicz, R. Hellwig, M. Uphoff, M. A. Öner, P. S. Deimel, Y.-Q. Zhang, C. A. Palma, F. Allegretti, M. Ruben, J. V. Barth and F. Klappenberger. "Proton tunneling-mediated on-surface hydroalkoxylation" *submitted*.

Acknowledgements

There are many people I wish to express my gratitude for their direct and indirect contribution to making this work possible.

First, I thank Prof. Dr. Johannes Barth for offering me a Ph.D. position in his research team and for his supervision, support, counseling on the results and correction notes.

Next, I thank Prof. Dr. Harald Brune for the opportunity to join his project collaboration, for making possible the synchrotron trips, and for his support and proofreading of manuscripts.

Many thanks to Prof. Dr. Florian Klappenberger for the helpful guidance and support throughout my doctorate study and for all the counseling and corrections for improving my writing skills. The discussions in his office were always very helpful, and I could learn much from his deep knowledge and research experience in STM and XS.

I express my gratitude to Prof. Dr. Peter Feulner for answering my technical and scientific questions, for sharing his ideas and immense knowledge in physics and electronic engineering, and for providing instructions on constructing electronic devices and UHV equipment.

For their assistance in technological problems, construction, and repair of laboratory equipment, support on computer-related issues, I want to thank Karl Eberle, Karl Kölbl, Reinhold Schneider, Hartmut Schlichting, and Max Glanz. I am very thankful to our secretary Viktoria Blaschek for handling official documents and for her helpful advice on organizational issues.

Furthermore, I thank Peter Weber for introducing me to the E20 group and for explaining me the experimental setup. Also, thanks to Dr. Saranyan Vijayaraghavan for the amusing coffee breaks making the day more comfortable.

Many thanks to my colleague Martin Uphoff for his large contribution to the STM experiments, technical ideas, and precise construction of new laboratory equipment (e.g., the urgently needed lanthanide evaporator and laboratory gas line). I also thank my colleague Tobias Paintner for his strong involvement in the STM experiments and help in the laboratory. Thank you both for the productive and cheerful time, the teamwork spirit, and for your commitment in our joint projects, some of which I presented here.

Special thanks go to Dr. Yiqi Zhang for the informative discussions and helpful comments on experimental results and computational simulations. Furthermore, thanks to Peter Hasch for the nice atmosphere and cheerful moments making the days in the office more comfortable. Also thanks to Liding Zhang, Dr. Anthoula Papageorgiou, Dr. Carlos-Andres Palma, and Prof. Dr. Willi Auwäter for the entertaining lunchtime meetings.

Many thanks to the beamline crew members Jan Dreiser, Katharina Diller, Romana Baltic, Aparajita Singha, Fabio Donati, and Stefano Rusponi for the warm welcome during my stay at the École Polytechnique Fédérale de Lausanne (EPFL) and X-Treme beamline at the Paul Scherrer Institute (PSI) in Zurich.

I would like to say "Muchas gracias" to Dr. David Écija and his beamline colleagues for the friendly welcome in his research group and the pleasant stay during the beamtime at the ALBA synchrotron in Barcelona.

For the synthesis of the investigated molecules, I thank the chemists Dr. Zhi Chen and Dr. Ping Du. Moreover, for performing DFT calculations on the organocobalt complex on Ag(111), my thanks go to Dr. Ari P. Seitsonen.

Furthermore, I thank Felix Bischoff, Jacob Ducke, Mateusz Paszkiewicz, Peter Deimel, Andreas Walz, Martin Schwarz, Borja Cirera, Alissa Wiengarten, Dr. Manuela Garnica, and Georg Michelitsch for the entertaining laboratory conversations, group meetings, and conference sessions. Moreover, thanks to all other persons within and outside of E20 group that supported me and I forgot to mention here. It has been a wonderful experience working with all of you.

My profound gratitude to Stephanye for supporting and accompanying me during the last 4 years. Lastly, I express my deepest thanks to my parents, brother, and two sisters for always being here for me.

Contents

Abbreviations

0D	Zero-dimensional
1D	One-dimensional
2D	Two-dimensional
3D	Three-dimensional
AFM	Atomic force microscopy
BE	Binding energy
CN-DETP	4,4″-diethynyl-[1,1′:4′,1″-terphenyl]-3,3″-dicarbonitril
CN-GDY	Cyano-functionalized 3-2 graphdiyne
CNT	Carbon nanotube
COF	Covalent organic framework
CVD	Chemical vapor deposition
DCD	Dewar–Chatt–Duncanson
DETP	4,4″-diethynyl-1,1′:4′,1″terphenyl
DFT	Density functional theory
DNA	Deoxyribonucleic acid
DOS	Density of states
Ext-TEB	1,3,5-tris(4-ethynylphenyl)benzene
FFT	Fast Fourier transform
FWHM	Full width at half maximum
GDY	Graphdiyne
GNR	Graphene nanoribbon
HEB	Hexakis((trimethylsilyl)ethynyl)benzene
IETS	Inelastic tunneling spectroscopy
JT	Joule–Thompson
JT-STM	Joule–Thompson scanning tunneling microscope
LDOS	Local density of states
LIA	Lock-in amplifier
LM	Lateral manipulation
LT	Low temperature
LT-STM	Low-temperature STM
LUMO	Lowest unoccupied molecular orbital

MA	Main axis
min	Minute
ML	Monolayer
MOCN	Metal–organic coordination network
MOF	Metal–organic framework
NEXAFS	Near-edge X-ray absorption fine structure
OFL	Open feedback loop
OMBE	Organic molecular beam epitaxy
PARI	Proton acceptor ring interaction
RT	Room temperature
STM	Scanning tunneling microscopy
STS	Scanning tunneling spectroscopy
sub-ML	Sub-monolayer
TEB	1,3,5-triethynyl-benzene
TEM	Transmission electron microscope
TMS	Trimethylsilyl
UHV	Ultra-high vacuum
WKB	Wentzel–Kramers–Brillouin
XPS	X-ray photoelectron spectroscopy
XS	X-ray spectroscopy

Chapter 1
Introduction

1.1 Motivation

Nature's virtuosity in creating a multitude of complex structures and functionalities has always enriched human reflection and creativity within different fields of knowledge. Remarkable technological achievements in materials science are often inspired by processes taking place at natural surfaces and interfaces, e.g., gecko-inspired dry adhesives [1], drag-reducing and self-cleaning coatings mimicking the texture of shark skin [2], waterproof sealings based on the hydrophobic texture of nasturtium leaves [3], etc. The structure and functionality within these systems requires the organization of few "simple" molecular units into complex architectures that support vital processes, such as oxygen transport in hemoglobin and light-harvesting in chlorophyll molecules.

Single crystal surfaces under ultra-high vacuum (UHV) conditions represent versatile platforms to study fundamental molecular processes under well-defined conditions. Scanning tunneling microscopy (STM) and photoelectron spectroscopy (XS) are key tools for surface and nanoscale science targeting the study of the physico-chemical properties of surface-adsorbed systems at the atomic and element-specific level, respectively, with key relevance for catalysis, sensor technology and molecular electronics [4]. The broad availability of custom-synthesized molecular building blocks has contributed to the success of supramolecular on-surface self-assembly [5, 6] in creating artificial organic architectures and metal-organic coordination networks (MOCN) with utmost precision and control.

The bottom-up approach of supramolecular chemistry has made possible the development of on-surface molecular engineering, i.e., a research field targeting the synthesis of metal-organic frameworks (MOF) and covalent organic frameworks (COF) through the carbon-metal and carbon–carbon bond formation between custom-designed monomers deposited on a well-defined surface. The resulting low-dimensional polymer structures are highly attractive for developing new concepts toward molecular electronics, i.e., light-emitting diodes, photovoltaic devices, carbon-based transistors, switches and rotors, transistors, sensors and storage units

R. Hellwig, *Alkyne-Based Nanostructures on Silver Substrates*, Springer Theses,
https://doi.org/10.1007/978-3-030-00997-7_1

Acetylenic derivatives represent a promising class of building blocks to extend the surface scientist's toolbox for the creation of low-dimensional materials with exquisite structure-function relationships. The alkyne's rich chemistry based on its unsaturated character, π-electron delocalization and linear geometry allows for a variety of interesting reaction mechanisms on surfaces. This thesis focuses on the on-surface reaction capabilities of alkyne-functionalized molecules confined on single-crystal silver substrates.

1.2 Elements of Nanoscience and Scanning Probe Microscopy

Already more than sixty years ago, Richard Feynman delineated in his famous speech "There's Plenty of Room at the Bottom", how the concerted assembly and manipulation of single atoms will revolutionize the computational performance and chemical synthesis routes in the future [7]. His visions are related to the essential aspects of nanotechnology, i.e., a term coined by Norio Taniguchi in 1974 [8] and used again by Kim Drexler in the context of molecular engineering and machinery, as published in his imaginative book "Engines of Creation" in 1986 [9, 10]. Meanwhile, the breakthrough discoveries, i.e., scanning tunneling microscopy (STM) and atomic force microscopy (AFM), invented and brought to perfection by Gerd Binnig and Heinrich Rohrer, have realized the scientific dream of visualizing atoms and molecular bonds on well-defined metallic and insulating surfaces [11–13]. At the crossroads of chemistry, biology, physics and engineering, these inventions caused a paradigm shift toward the study on nanoscale objects. STM permitting the atomic-level characterization of single-crystal samples immediately spurred new scientific achievements including the structural elucidation of the Si(111) 7×7 reconstruction [14] and the Au(111) chevron reconstruction [15, 16]. Well-defined surfaces were then used as two-dimensional (2D) platforms to study single atoms and molecules up to the complexity of deoxyribonucleic acid (DNA) [17], whose structural and electronic properties dramatically deviate from their gas-phase character. The potential of STM in creating artificial structures was shown by Eigler et al., who positioned single atoms to artificially create nano-objects in a step-by-step fashion, e.g., the acronym "IBM" made up of Xe atoms on Ni(111) [18]. Subsequently, Crommie et al. created a quantum resonator comprised of single Fe atoms on Cu(111) [19]. By positioning one Co atom at one of the two elleptical foci, Manoharan et al. showed by scanning tunneling spectroscopy (STS) that the Kondo effect of Co/Cu(111) in one focus of the ellipse is created at the other focus by scattering and focusing of surface state electrons (quantum mirage of a Kondo resonance) [20]. Based on the concerted manipulation on single atoms and molecules, STS measurements have unraveled fascinating phenomena including single-atom switching of charge states [21], single-molecule transistors gated by single atoms [22], photoexcitation and fluorescence spectroscopy [23, 24], vibrational spectroscopy [25], chemical reactions triggered by the STM tip [26], molecular switches and rotors [27, 28], molecular superconductivity [29], magnetism within 2D graphene [30], atom-precise fabrication of magnetic structures [31] and single-atom spin relaxation [32].

Self-assembly on Surfaces

While the atom-by-atom arrangement via the STM tip is time-consuming and limited to a few nanostructures of small size, extended regular nanoarchitectures comprised of $\sim 10^{12}$ units per cm^2 are instantaneously created via the self-assembly of vacuum-deposited adsorbates on single-crystal surfaces. For example, the self-organized growth of silver atoms on Pt(111) toward fractal, dentritic and equally spaced islands on strain-relief patterns, as reported by Brune et al. [33, 34], or a surface state-mediated hexagonal superlattice of cerium adatoms on Ag(111) [35]. These adsorbate structures represent model systems with catalytic, electronic and magnetic properties that can dramatically scale with the adsorbate size [34]. While the organization of metallic aggregates is influenced by epitaxial parameters such as the mismatch between host and adsorbate lattice constants, supramolecular self-assembly is programmed via the functional design of the molecular species. Accordingly, chemically synthesized molecular units are used to construct nanoarchitectures through complementary recognition and lock-and-key principles originating from the molecule's functionalization. A prerequisite for the creation of highly organized organic [36] and metal-organic surface-confined structures [37] is the presence of reversible and non-covalent intermolecular interactions enabling correction mechanisms for eliminating structural defects. From the STM-based exploration of metal-molecule interfaces, fundamental principles of supramolecular chemistry and hierarchical recognition phenomena within biomoleculecules and living organisms can be mimiced in a controlled way on surfaces under well-defined conditions (low temperature (LT) and UHV), thereby enabling approaches toward bio-inspired interfaces [38].

Both the self-assembly and artificial STM tip-assisted positioning of adsorbates on surfaces represent an atom-precise bottom-up strategy to fabricate low-dimensional nanostructures. It is complementary to top-down techniques that use photolithographic masks to create periodic templates via the removal of material from a bulk-like structure. For a profound analysis of the geometrical, conformational and chemical aspects of on-surface synthesized patterns, the synergy between local STM microscopy, space-averaging XS, together with theoretical modeling has proved to be very useful [39].

1.3 Carbon Chemistry

Carbon is the most versatile element on earth making up the complexity of organic matter and life. The rich chemistry of carbon structures is based on its large number of oxidation numbers ranging from –4 in methane (CH_4) to +4 in carbondioxide (CO_2) (within the oxidation number formalism [40]) and its capability to bond with itself, e.g., within the cyclic, conjugated and aromatic structure of benzene (C_6H_6), discovered in 1865 by August Kékulé. The vast field of organic chemistry stems from the reactivity and structural diversity of carbon bonds giving rise to different shapes, e.g., rings, chains, spheres, tubes, 2D sheets and 3D materials, as explained by the concepts of orbital hybridization developed by Linus Pauling [41]. The chemical versatily of carbon gives rise to fourfold covalent bonding and tetrahedral arrangement within the sp^3-hybridized structure of diamond, threefold bonding and trigonal planar arrangement within the sp^2-hybridized layers of graphite and twofold bonding and straight-line arrangement within sp^1-hybridized polyyne oligomers of different length (see Fig. 1.1). The smallest hydrocarbon molecule according to each hybridization class is represented by methane (one C atom with four C–H single bonds), ethylene (C=C double bond with four C–H single bonds) and acetylene (C≡C triple bond with two C–H single bonds).

The multifaceted aspects of carbon chemistry also include the field of organometallic chemistry, as demonstrated in 1890 by Ludwig Mond who used carbonyl molecules (C≡O) to purify the 3d metal Ni (Mond process), thereby establishing one of the first organometallic complexes, namely $Ni(CO)_4$ [42]. It was shown by Nikolaev et al. that a related metallocarbonyl complex ($Fe(CO)_5$) serves as a catalyst for the conversion of CO gas into single-walled carbon nanotubes (CNT) with a diameter of ~0.7 nm [43].

1.3.1 Sp²-Hybridized Carbon Allotropes

Diverse sp^2-hybridized carbon allotropes with well-defined structural properties down to the atomic level are hitherto known. Besides naturally occuring diamond and graphite solids, prominent examples are zero-dimensional (0D) fullerenes (C_{60} discovered in 1985), one-dimensional (1D) CNT representing one seamlessly coiled graphite layer (discovered in 1991) and 2D graphene (explored in 2004) (see Fig. 1.1d). They can be artificially synthesized through the sublimation of graphite clusters via arc discharge and laser ablation techniques as well as chemical vapor deposition (CVD) (continuous gas-flow and thermal decomposition of precursor molecules), whereby the selectivity toward the targeted product can be controlled through the size and structure of catalytic interfaces (e.g., 3d metal clusters and surfaces) [43, 46]. The key material graphene, i.e., a one-atom thick sheet of sp^2-hybridized C atoms is famous for its extraordinary mechanical, thermal and optical properties raising high hopes toward new materials and transistor devices [47, 48].

However, one severe obstacle toward applications in nanoelectronics is the absence of a bandgap within graphene. Hence, there is a high interest in creating related carbon allotropes of low dimensionality (1D and 2D) and of different/mixed hybridization states that allow wide bandgap tuning.

1.3.2 Sp-Hybridized and Mixed Sp-Sp^2 Materials

Among the many hypothetical structures containing C atoms of different hybridization states, including 0D, 1D and open-porous 2D allotropes (Fig. 1.1), theoretical simulations predict exciting properties for sp-hybridized and mixed sp/sp^2 materials (for example sp–sp^2-graphyne [49] shown in Fig. 1.1d). Due to the existence of delocalized π electrons determining through-bond transport properties, the structural diversity of these materials goes in hand with their high potential toward carbon-based molecular electronics (e.g., transistors, semiconductors, metallic molecular wires). Their physicochemical properties can be further refined by the incorpora-

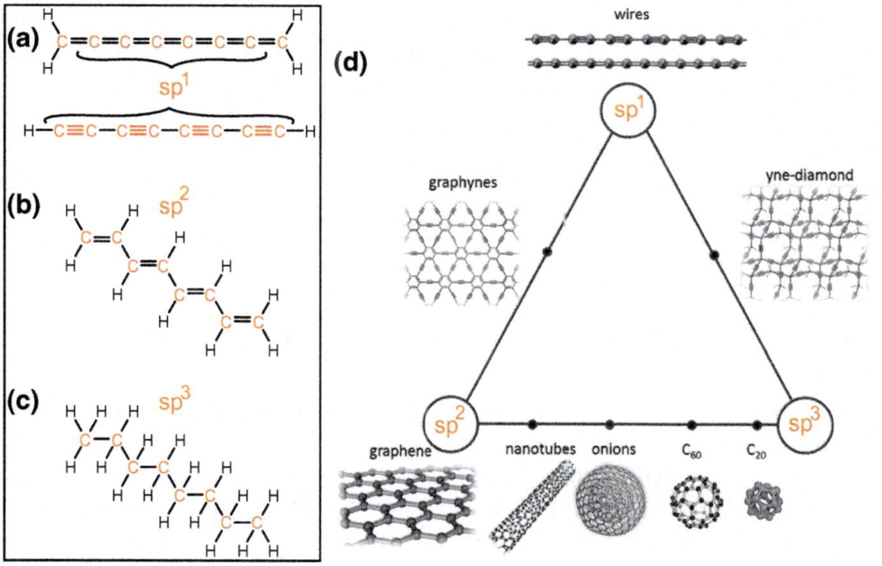

Fig. 1.1 Carbon-rich structures subdivided according to their hybridization character. **a–c** Homologous series of hydrocarbon chains, each displaying eight C atoms. **a** Linear chains from sp-hybridization: cumulene $(=C=C=)_n$ and polyyne $(-C\equiv C-)_n$ (here: n = 4) [44]. Triple bonds (alkynes) are colored in orange. Terminal C atoms of cumulene are sp^2-hybridized. **b, c** Zig-zag chains from sp^2-hybridization within polyacetylene (here: tetraene derivative) and sp^3-hybridization within alkane polymer (here: octane). **d** Carbon allotropes with different hybridization characteristics. Image in (**d**) taken from [45]

tion of redox-active metal centers, thereby tuning their eletronic, luminescent and magnetic properties [46, 47].

The sp-hybridized character of the triple bond within acetylene allows the formation of a single-atom thick carbon chain of equally spaced C atoms, i.e., a linear cumulene wire $C_{2n}H_2$ of length n $\geq$ 2 (Fig. 1.1a). Such sp-unsaturated polymer wires reveal interesting electronic, optical and conformational properties different from their sp^2- and sp^3-hybridized counterparts polyacetylene (C_nH_{n+2}) and alkane (C_nH_{2n+2}) (see Fig. 1.1b, c) [44]. Strain within the carbon wire can lead to bond length variations causing a transition from cumulene to polyyne, i.e., a carbon chain of alternating triple and single bonds (Fig. 1.1a). The Peierls distortion [50] associated with this structural change leads to a metal-semiconductor transition with tunable bandgap [51, 52]. Moreover, the linear structure of polyyne is interesting for studying the mechanical bending capabilities of single-atom thick chains [53].

Due to their cylindrical π-electron delocalization around the main axis, C(sp) structures reveal unique electron characteristics that can be combined with sp^2-hybridized structures, e.g., a polyyne wire enclosed by a CNT (peapod structure) [54] or bound to a graphene nanoribbon (GNR) [51]. Such hybrid structures extend the toolbox for carbon-based materials with individual electron characteristics.

As a consequence of the high reactivity of C(sp) atoms, the synthesis and characterization of these molecular rod-like carbyne structures still remains challenging. Various reports claim the creation of single-atom thin carbon chains based on the laser ablation of graphite particles in a solution environment [55], the etching of a graphene layer via the electron beam of a transmission electron microscope (TEM) [56] and the capture of C atoms through graphene-supported Pt atoms serving as nucleation sites for the carbon chains [44]. However, an atomically precise characterization of their structural properties within isolated polymer strands is still missing.

1.3.3 Covalent Coupling of Terminal Alkyne Monomers

Low-dimensional carbon-rich scaffolds with sp-hybridized motifs can be created through the linkage of ethynyl-functionalized monomers [45]. As shown at the very top of Fig. 1.2a, the dehydrogenative coupling of two terminal alkyne monomers leads to a covalent dimer expressing a diyne motif (similar structure as polyyne (Fig. 1.1a), but for $n = 2$). The latter connects two (aromatic) molecular backbones depicted as residual groups (R) in Fig. 1.2a, b. The schematic illustration in Fig. 1.2a shows two approaches for the terminal alkyne coupling of monomers, i.e., in-solution reactions versus on-surface reactions. The knowledge obtained from solution-mediated alkyne chemistry is used to create low-dimensional materials through the on-surface carbon–carbon bond formation between UHV-deposited molecules.

Figure 1.2b reveals solution-based terminal alkyne coupling schemes catalyzed by 3d metal atoms. Within the shown reaction pathways toward diyne dimers, organometallic bonding structures represent the intermediates for covalent coupling. The first pioneering work on acetylenic coupling dates back to 1869, when the

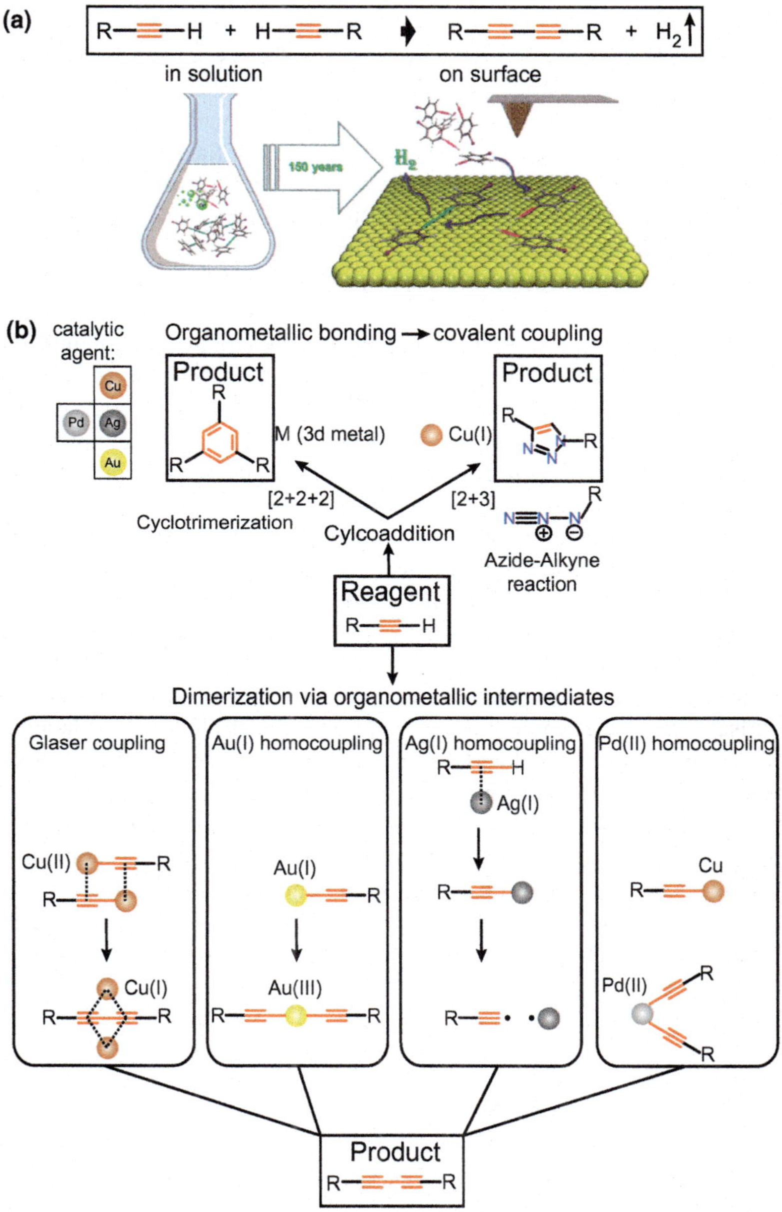

Fig. 1.2 Alkyne-based covalent coupling (see scheme in the top part of **a**). **a** In-solution coupling versus on-surface coupling of terminal alkynes on metal substrates (right) (sketch from [58]). **b** Solution-based coupling reactions catalyzed by noble metal atoms, i.e., [2+2+2] and [2+3] cycloaddition as well as Glaser coupling and homocoupling mediated by noble metal atoms. Organometallic intermediates for dimerization are depicted

chemist Carl Glaser discovered the oxidative dimerization of copper phenylacetylide toward diphenyldiacetylene. The reaction was further optimized by Eglinton and Galbraith, who chose specific concentrations of cupric salts in methanolic pyridine as oxidant, and by Hay, who combined Cu(I) chloride with the ligand N,N,N′N′-tetramethylethylenediamine to increase the solubility of the reactants participating in oxidative coupling [57]. While a complete chemical picture of the Glaser coupling is still missing, Bohlmann et al. suggested a dinuclear Cu(II) intermediate that transforms into the targeted structure before releasing bound Cu(II) ions. A different reaction route is accomplished by firstly deprotonating the alkyne and establishing an (alkynyl)cuprate complex through the addition of copper(I) iodide. Afterward, Pd(II) is introduced to transform the compound into diethynyl-palladium(II), which disintegrates into a straight sp-hybridized diyne and Pd(0). The intermediate of Au(I)-homocoupling is a linear C(sp)–Au–C(sp) species, which directly forms the targeted product after the release of the Au ion. Alkyne-based organosilver complexes related to Ag-mediated homocoupling are scarcely explored and applied in organosynthesis processes. The first reaction step resulting in a π complex requires KOAc to deprotonate the alkyne and establish a Ag-alkynyl compound. Interestingly, the Ag-homocoupling follows a radical mechanism, in which the homolysis of C–Ag bonds delivers alkynyl radicals that recombine via the formation of 1,3-diynes.

The versatility of terminal alkyne reactions is not limited to twofold homocoupling, but also includes [2+2+2] and [2+3] cycloaddition reactions. In the former case, transition metal atoms catalyze the transformation of three terminal alkynes into a trisubstituted benzene ring (upper part of Fig. 1.2b) [58]. In the latter case, a Cu(I)-catalyzed dipolar coupling between azide and alkyne moieties results in the cycloaddition toward a five-membered 1,2,3-triazole ring. This "click reaction" [59] even proceeds in the absence of a metal catalyst, however with significantly reduced reaction rates and regioselectivy (more byproducts) [60]. A further terminal alkyne coupling mechanism is based on Sonogashira coupling, which is a cross-coupling reaction between aryl halides and alkynes resulting in a R–C≡C–R linkage instead of a 1,3-diyne motif. The reaction can be activated by a Pd-triphenylphosphine complex [61] or a AgI catalyst [62].

1.3.4 Metal Alkynyl Complexes

Within the reaction types toward terminal alkyne dimerization shown in Fig. 1.2b, organometallic structures represent intermediates for the conversion into covalent diyne motifs after the release of incorporated metal ions. However, there exist also reaction routes leading to metal-carbon bonding configurations, which do not lead to covalent carbon–carbon bond formation, i.e., organometallic bonding is stably established. Figure 1.3a reveals a metal polyyne polymer, i.e., a 1D wire of alternating metal-carbon and C≡C bonds (cf. polyyne wire in Fig. 1.1a). Such alkyne-based metallopolymers incorporating transition metal atoms (M) along the chain axis, e.g., Pt-acetylides [63], can exhibit interesting optoelectronic properties

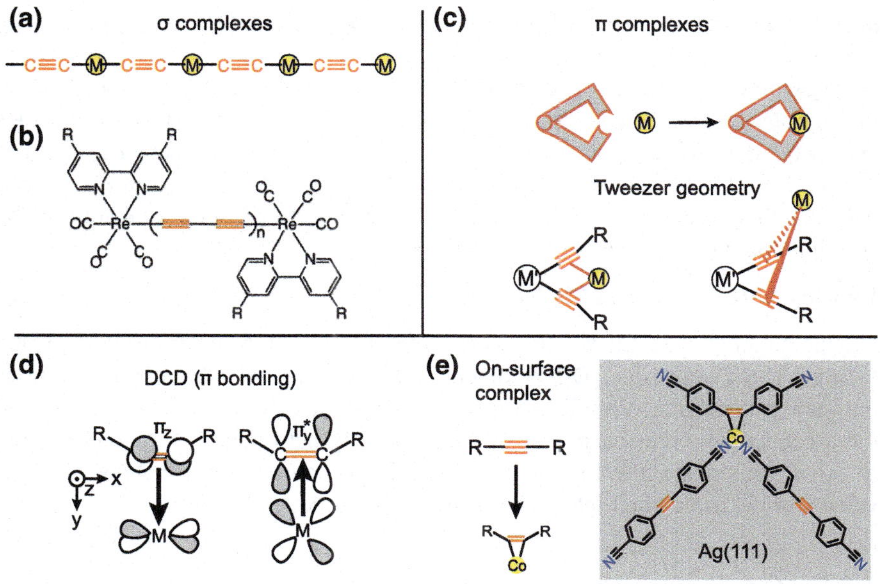

Fig. 1.3 σ-like (**a**, **b**) and π-like (**c**) organometallic bonding within metal-alkynyl complexes. M and M' indicate metal centers; N and Co refer to nitrogen and cobalt atoms. **a** Metal polyyne polymer. **b** Dinuclear alkynylrhenium(I) complex (from [76]) **c** Two alkynes expressing tweezer-like bonding within metal alkynyl complex. Trapped metal atom M lies within (above) the plane spanned by the two alkynes bound to M'. **d** DCD model explaining bonding between π orbitals and d orbitals of metal M. **e** On-surface formation of an organocobalt complex comprised of three monomers expressing a Co-alkynyl motif stabilzed by metal-ligand bonding

due to the alkyne's π conjugation within its linear geometry combined with an increased spin-orbit coupling attributed to the metal atoms [64]. In addition to their potential as light-emitting diodes, metal alkynyl complexes are interesting candidates for photovoltaic applications [65] and also reveal attractive nonlinear optical properties sensitive to the oxidation states at the metal centers [66]. Dinuclear compounds containing two transition metal atoms coordinated to alkynes (Fig. 1.3b) can exhibit electronic coupling and electron delocalization through C(sp) atoms [67], coupling of multiple spin centers [68] and mixed-valence properties interesting for electrochemical switching applications [66, 69]. Furthermore, the control on metal-metal interactions within polynuclear alkynyl complexes could be used to refine their photochemical and photophysical properties [70]. Metallopolymers containing multiple metal atoms can be moreover used for molecular imaging purposes [71] and are relevant as macromolecular catalysts mimicking the biological function of enzymes [72].

Besides σ bonding, the alkyne's π orbitals can get involved in π bonding with metal atoms. Figure 1.3c displays a π complex characterized through the tweezer-like arrangement of two alkyne units mediating site-selective interactions between π orbitals and the orbitals of M. Since C(sp) atoms exhibit perpendicular π orbitals, the trapped atom within this host-guest π complex [73] might be located within or outside of the molecular plane, as shown in the left and right part of Fig. 1.3c. Thus, alkyne-based molecules permit the synthesis of organometallic architectures with large structural diversity and tunability.

Organometallic motifs different from solution-based chemistry can be created through the on-surface coupling (explained below) of surface-adsorbed molecules and metal atoms on an inert metal substrate. Figure 1.3e shows a scheme of an on-surface formed organocobalt π complex on top of the Ag(111) surface: a trimer held together through C(sp)–Co interactions, while the Co atom is stabilized by coordination with two cyano groups [74].

Dewar-Chatt-Duncanson Model

Organometallic π complexes are usually described within the framework of the Dewar-Chatt-Duncanson (DCD) model [75], which considers the interactions between molecular π orbitals and metallic d orbitals (Fig. 1.3d). In the context of the organocobalt complex in Fig. 1.3e, the synergy between $\pi \rightarrow d$ donation and $\pi^* \leftarrow d$ backdonation from molecular orbitals can cause bond stretching and conversion of the triple bond into a double bond (Fig. 1.3d, e).

1.3.5 On-Surface Polymerization

The on-surface coupling of monomers deposited under UHV conditions on well-defined metal substrates exhibits advantages over coupling reactions in 3D solution environments. Firstly, the diffusion of surface-confined monomers is restricted to two dimensions (cf. illustration in Fig. 1.2a), which is a necessary prerequisite for the targeted synthesis of low-dimensional covalent organic frameworks (COF). In addition, no extra functional groups or ligands are necessary in order to enhance monomer solubility [58]. While the solution is comprised of solvent, molecular reagents, catalytic agents (transition metal atoms), oxidants, reducing agents, molecular intermediates and products, the metal surface represents a catalytic interface solely confining monomers, intermediates and coupling products.

The solvent-free deposition of functionalized monomers on metal substrates bestows versatile reaction routes including Ullmann-type aryl-aryl coupling [77, 78], (poly)condensation reactions [79–82], transition metal-assisted polymerization [83] and cyclotrimerization of acetyls [84]. With the support from theory calculations delivering details on chemical transformations, intermediate states, activation barriers and reaction pathways, on-surface chemistry targets the atom-precise engineering of only-carbon scaffolds with tunable structural and electronic properties. The targeted 2D open-porous polymers are highly attractive owing to tunable electronic properties [85, 86] and technological applicability [77, 79, 87, 88] as molecular sieves, energy storage units [89], molecular nanocircuits and host-guest architectures. The creation of regular covalent nanostructures through the coupling of monomers requires the control of molecule-surface and non-reversible intermolecular interactions, which are governed by reaction kinetics and entropy. Low-temperature STM (LT-STM) is exquisitely suited to perform atom-precise imaging, local manipulations and spectroscopic measurements on monomers, intermediates and final structures.

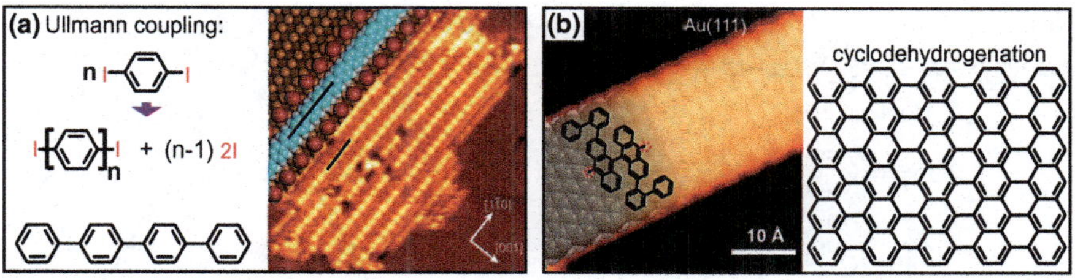

Fig. 1.4 **a** 1D polyphenylene nanowires obtained by Ullmann coupling of 1,3-diiodobenzene on Cu(110) (STM image from [90]). Dehalogenative dissocation preceding intermolecular coupling at 500 K gives rise to iodine byproducts (red spheres). **b** GNR created via dehalogenative coupling of 10,10′-dibromo-9,9′-bianthracene precursors on Au(111) and subsequent cyclodehydrogenation at 673 K (STM image from [91]). Adjoined carbon lattice model represents arm-chair GNR

Ullmann Reaction on Surfaces

A promising strategy to create covalent graphene-like structures is based on the on-surface aryl-aryl coupling (Ullmann reaction). As shown in Fig. 1.4a, a ditopic halogen-functionalized monomer confined on Cu(110) is used to construct 1D polymer strands. The dissociation of terminal halogen atoms precedes the coupling of the end groups via radical recombination. The coupling of n 1,3-diiodobenzene molecules corresponds to the cleavage of $2 \times (n-1)$ iodine atoms which remain adsorbed on the surface (next to the on-surface formed polyphenylene wire in Fig. 1.4a) [90]. Targeting the growth of GNR of defined width, which determines its electronic bandgap, Chen et al. employed a specific monomer design (Fig. 1.4b). After dehalogenation and bond formation, heat-induced cyclodehydrogenation on Au(111) (at 673 K) leads to a flat GNR with hydrogen-terminated zigzag edges [91]. Because of the atom-precise control over charge-carrier mobilities, electron delocalization and bandgap alignment, on-surface coupling could lay the foundations for carbon-based field-effect transistors replacing current silicon-based technologies [92].

Terminal Alkyne Reactions on Surfaces

In contrast to Ullmann coupling, terminal alkyne homocoupling (known from solution chemistry, cf. Fig. 1.2b) on metal surfaces represents a clean reaction only releasing volatile hydrogen, which already desorbs at substrate temperatures of $T \sim 200$ K on Ag(111) (see homocoupling scheme in Fig. 1.5a). While the inertness of saturated terminal alkyl groups (C(sp^3)) requires high substrate temperatures to create organometallic carbon-metal-carbon bonds [93] and covalent carbon–carbon bonds [94], the highly unsaturated character of the alkyne leads to low activation barriers toward on-surface polymerization via C(sp)-metal-C(sp) linkage [95, 96] and covalent C–C coupling [97, 98].

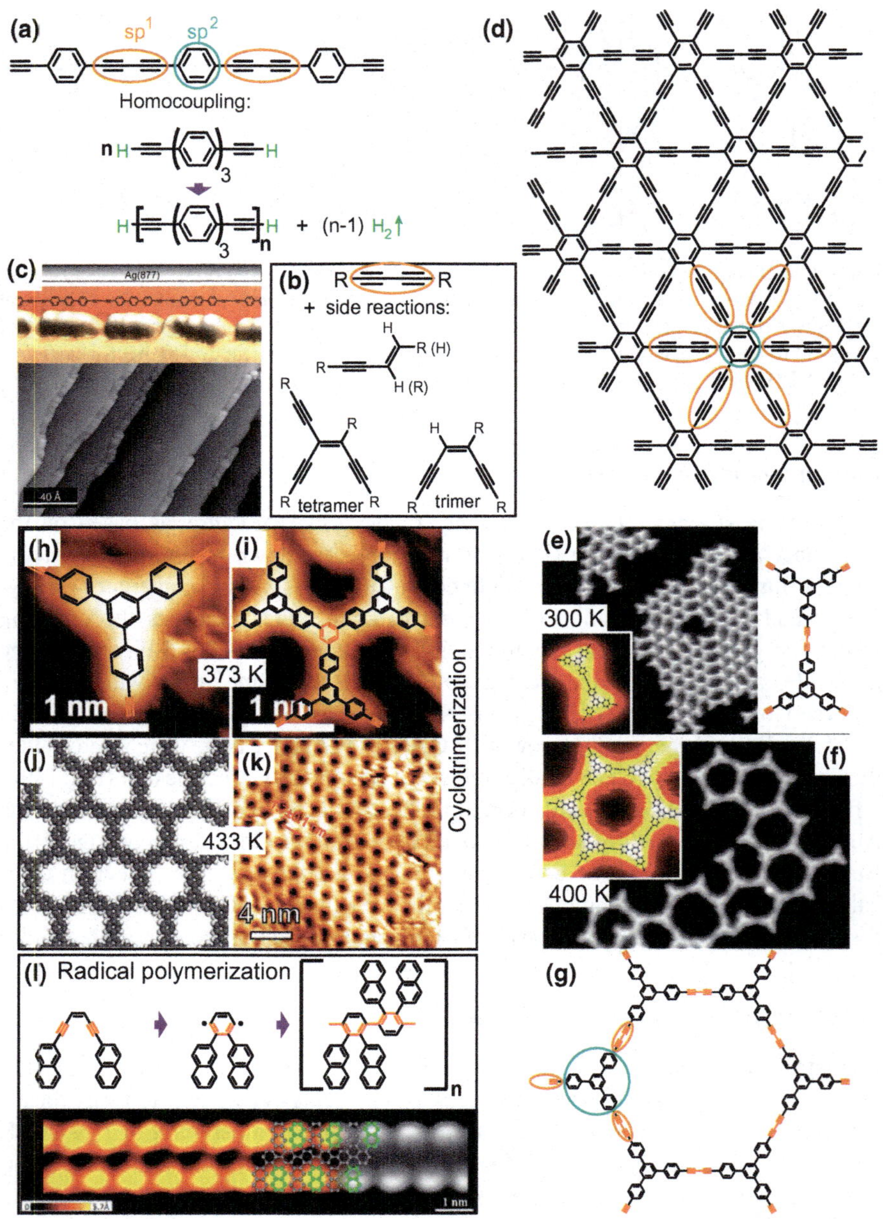

Fig. 1.5 Alkyne-based on-surface polymerization. **a** Model structure of GDY nanowire. Terminal alkyne homocoupling (**a**, **b**) of a diethynyl derivative on Ag(877) leads to extended GDY nanowires (**c**) (STM image from Ref. [98]). **d** Model structure of 2D GDY. **e–g** Homocoupling of Ext-TEB on Ag(111): formation of dimers (**e**) and porous structures (**f**, **g**) (STM images from Ref. [97]). **h–k** Cyclotrimerization of Ext-TEB on Au(111): from monomers (**h**) to trimers (**i**) and extended scaffolds (**k**) (STM images from [102]). **l** Bergman cyclization of 1,6-di-2-naphthylhex-3-ene-1, 5-diyne on Cu(110) results in polyphenylene nanowires (STM image from [103])

Terminal alkyne homocoupling allows the fabrication of extended graphdiyne (GDY) structures [99], i.e., hydrocarbon architectures representing an alternating sequence of the linear diyne unit and the molecular backbone. Models for 1D and 2D GDY are shown in Fig. 1.5a, d, whereby orange and cyan outlines enclose sp- and sp^2-unsaturated regions of the polymer. In contrast to graphene, 2D GDY expresses a mesoporous structure with a non-zero electronic bandgap between Dirac cones, which can be modulated through external strain on the polymer [100]. Due to sp-hybridization, there is a perpendicular π system, which may represent multiple binding sites for metal-doping [101]. Based on the high π conjugation and tunable properties of these sp–sp^2-hybrids, their predicted electronic, mechanical and physico-chemical properties may lead to new applications, such as molecular sieves, H_2 and Li ion storage units, carbon-based semiconductors, field-effect transistors, chemical sensors and photovoltaic devices [100, 101].

Although the methodology to fabricate GDY-like structures seems to be straightforward, experimental results are still rudimentary due to difficulties in achieving the required chemoselectivity for homocoupling toward the targeted diyne linking motifs and in inhibiting byproducts from side reactions. As shown in Fig. 1.5b, the latter are twofold vinyl-ethynyl regiomers (head-to-tail and head-to-head enynes), trimerization, tetramerization and Ag-acetylide motifs [58]. To reduce the amount of byproducts, Cirera et al. employed the ditopic alkyne linker 4,4″-diethynyl-1,1′:4′,1″-terphenyl (DETP) on the vicinal Ag(877) surface, which leads to a significantly higher chemoselectivity toward the targeted linear linkage mode related to homocoupling, as compared to the less corrugated Ag(111) surface (Fig. 1.5a–c). The extended GDY nanowires are aligned along the step edges of Ag(887) (Fig. 1.5c). Zhang et al. reported homocoupling on Ag(111) for the tritopic molecule 1,3,5-tris(4-ethynylphenyl)benzene (Ext-TEB), which couples to dimers (at ~300 K) and oligomeric patches, i.e., closed polygon motifs within irregular hydrocarbon nanomeshes (at ~400 K) (Fig. 1.5e, f) [97]. The absence of long-range order is due to side reactions causing the coexistence of polygons of different symmetry. From the model for the sixfold polygons (Fig. 1.5g), one recognizes the tritopic backbone of Ext-TEB (cyan) and the diyne linkage (orange).

Targeting the synthesis of GDY-like materials, Sun et al. employed a different type of terminal alkyne reaction, i.e., dehalogenative coupling of terminal alkynyl bromides on the Au(111) surface. In this case, a 2D organometallic C(sp)–Au–C(sp) network is reported as an intermediate structure [96], which indicates the role of Au atoms in catalyzing C–C bond formation, in analogy to the solution-based C(sp)–Au–C(sp) intermediates in Fig. 1.2b.

For comparison, the related precursor Ext-TEB (with terminal hydrogen instead of halogen atoms) follows a cyclotrimerization pathway on Au(111), which differs from homocoupling on Ag(111). This reaction leads to a purely sp^2-hybridized polymer network (Fig. 1.5h–k) instead of a sp-hybridized structure. Figure 1.5i shows a trimer exhibiting a substituted benzene compound (orange), which emanates from the coupling of three terminal alkynes (orange).

A further alkyne coupling mechanism toward conjugated polymers is based on the Bergman cyclization of two internal alkynes next to a central double bond (Fig. 1.5l).

During this byproduct-free reaction, each monomer transforms into a diradical intermediate via a ring closure reaction. Radical recombination promotes the covalent coupling toward extended polyphenylene wires after sample annealing on Cu(110) [103].

1.4 Thesis Outline

The next chapter presents the basic theory of the employed experimental techniques and describes the utilized STM setup. In the main part, the STM results on the studied surface-adsorbed molecular systems are presented. The Chaps. 3 and 5 focus on the on-surface synthesis and STM characterization of alkyne-based organometallic structures. Specifically, the de novo synthesized alkyne linker 4,4″-diethynyl-[1,1′:4′,1″-terphenyl]-3,3″-dicarbonitril (CN-DETP) is used to steer the on-surface creation of cyano-functionalized Ag-bis-acetylide chains based on the coupling of terminal alkynes with Ag adatoms from the Ag(110) surface. The STM results are supported by complementary XPS and NEXAFS results revealing electronic and conformational properties (Chap. 3). Polynuclear 3d and 4f metal alkynyl complexes are created through the co-deposition of the molecule hexakis((trimethylsilyl)ethynyl)benzene (HEB) and metal atoms on the Ag(111) surface (Chap. 5). The experimental results in the Chaps. 4 and 6 delineate the synthesis of covalent structures in response to heat-induced and lanthanide-catalyzed terminal alkyne reactions on Ag(111). In the context of CN-DETP, the Ag(111) surface catalyzes the formation of cyano-functionalized GDY nanowires through terminal alkyne homocoupling. The high chemoselectivity toward covalent C(sp)–C(sp) bond formation on Ag(111) contrasts the high stability of carbon–silver bonds on Ag(110) (Chap. 4). Contrary to heat-assisted on-surface coupling, Chap. 6 presents a lanthanide-catalyzed alkyne cyclotrimerization reaction realized through the low-temperature dosage of Ho on Ag(111). Accordingly, threefold coupling of the employed tritopic alkyne monomer Ext-TEB results in the formation of open-porous oligomer structures. The final chapter summarizes the results of this thesis and delivers perspectives for future research projects.

References

1. Hawkes EW, Eason EV, Christensen DL, Cutkosky MR (2014) Human climbing with efficiently scaled gecko-inspired dry adhesives. J R Soc Interface 12
2. Ball P (1999) Engineering shark skin and other solutions. Nature 400:507–509
3. Bird JC, Dhiman R, Kwon H-M, Varanasi KK (2013) Reducing the contact time of a bouncing drop. Nature 503:385–388
4. Besenbacher F (1996) Scanning tunnelling microscopy studies of metal surfaces. Rep Prog Phys 59:1737
5. Cram DJ (1988) The design of molecular hosts, guests, and their complexes (nobel lecture). Angew Chem Int Ed 27:1009–1020

6. Barth JV (2007) Molecular architectonic on metal surfaces. Annu Rev Phys Chem 58:375–407
7. Feynman RP (1960) There's plenty of room at the bottom. Eng Sci 23:22–36
8. Taniguchi N et al (1974) On the basic concept of nanotechnology. In: Proceedings of the International Conference on Production Engineering Tokyo, Part II, Japan Society of Precision Engineering, pp 18–23
9. Drexler KE (1981) Molecular engineering: an approach to the development of general capabilities for molecular manipulation. Proc Natl Acad Sci USA 78:5275–5278
10. Drexler KE, Minsky M (1990) Engines of creation, Fourth Estate London
11. Binnig G, Rohrer H, Gerber C, Weibel E (1982) Tunneling through a controllable vacuum gap. Appl Phys Lett 40:178–180
12. Binnig G, Rohrer H, Gerber C, Weibel E (1982) Surface studies by scanning tunneling microscopy. Phys Rev Lett 49:57–61
13. Binnig G, Quate CF, Gerber C (1986) Atomic force microscope. Phys Rev Lett 56:930–933
14. Binnig G, Rohrer H, Gerber C, Weibel E (1983) 7×7 reconstruction on Si(111) resolved in real space. Phys Rev Lett 50:120–123
15. Wöll C, Chiang S, Wilson RJ, Lippel PH (1989) Determination of atom positions at stacking-fault dislocations on Au(111) by scanning tunneling microscopy. Phys Rev B 39:7988–7991
16. Barth JV, Brune H, Ertl G, Behm RJ (1990) Scanning tunneling microscopy observations on the reconstructed Au(111) surface: atomic structure, long-range superstructure, rotational domains, and surface defects. Phys Rev B 42:9307–9318
17. Beebe T, Wilson T, Ogletree F, Katz J, Balhorn R (1989) Direct observation of native DNA structures with the scanning tunneling microscope. Science 243:370–372
18. Eigler DM, Schweizer EK (1990) Positioning single atoms with a scanning tunneling microscope. Nature 344:524–526
19. Crommie MF, Lutz CP, Eigler DM (1993) Confinement of electrons to quantum corrals on a metal surface. Science 262:218–220
20. Manoharan HC, Lutz CP, Eigler DM (2000) Quantum mirages formed by coherent projection of electronic structure. Nature 403:512–515
21. Repp J, Meyer G, Olsson FE, Persson M (2004) Controlling the charge state of individual gold adatoms. Science 305:493–495
22. Martínez-Blanco J, Nacci C, Erwin SC, Kanisawa K, Locane E, Thomas M, von Oppen F, Brouwer PW, Fölsch S (2015) Gating a single-molecule transistor with individual atoms. Nat. Phys. 11:640–644
23. Wu SW, Ogawa N, Ho W (2006) Atomic-scale coupling of photons to single-molecule junctions. Science 312:1362–1365
24. Qiu XH, Nazin GV, Ho W (2003) Vibrationally resolved fluorescence excited with submolecular precision. Science 299:542–546
25. Stipe BC, Rezaei MA, Ho W (1998) Single-molecule vibrational spectroscopy and microscopy. Science 280:1732–1735
26. Hla S-W, Meyer G, Rieder K-H (2001) Inducing single-molecule chemical reactions with a UHV-STM: a new dimension for nano-science and technology. Chem Phys Chem 2:361–366
27. Auwärter W, Seufert K, Bischoff F, Ecija D, Vijayaraghavan S, Joshi S, Klappenberger F, Samudrala N, Barth JV (2012) A surface-anchored molecular four-level conductance switch based on single proton transfer. Nat Nanotechnol 7:41–46
28. Perera UGE, Ample F, Kersell H, Zhang Y, Vives G, Echeverria J, Grisolia M, Rapenne G, Joachim C, Hla S-W (2013) Controlled clockwise and anticlockwise rotational switching of a molecular motor. Nat Nanotechnol 8:46–51
29. Clark K, Hassanien A, Khan S, Braun K-F, Tanaka H, Hla S-W (2010) Superconductivity in just four pairs of (BETS)2GaCl4 molecules. Nat Nanotechnol 5:261–265
30. González-Herrero H, Gómez-Rodríguez JM, Mallet P, Moaied M, Palacios JJ, Salgado C, Ugeda MM, Veuillen J-Y, Yndurain F, Brihuega I (2016) Atomic-scale control of graphene magnetism by using hydrogen atoms. Science 352:437
31. Khajetoorians AA, Wiebe J, Chilian B, Lounis S, Blugel S, Wiesendanger R (2012) Atom-by-atom engineering and magnetometry of tailored nanomagnets. Nat Phys 8:497–503

32. Loth S, Etzkorn M, Lutz CP, Eigler DM, Heinrich AJ (2010) Measurement of fast electron spin relaxation times with atomic resolution. Science 329:1628–1630
33. Brune H, Romainczyk C, Roder H, Kern K (1994) Mechanism of the transition from fractal to dendritic growth of surface aggregates. Nature 369:469–471
34. Brune H, Giovannini M, Bromann K, Kern K (1998) Self-organized growth of nanostructure arrays on strain-relief patterns. Nature 394:451–453
35. Silly F, Pivetta M, Ternes M, Patthey F, Pelz JP, Schneider W-D (2004) Creation of an atomic superlattice by immersing metallic adatoms in a two-dimensional electron sea. Phys Rev Lett 92:016101
36. Pawin G, Wong KL, Kwon K-Y, Bartels L (2006) A homomolecular porous network at a Cu(111) surface. Science 313:961–962
37. Barth JV (2009) Fresh perspectives for surface coordination chemistry. Surf Sci 603:1533–1541
38. Abb S, Harnau L, Gutzler R, Rauschenbach S, Kern K (2016) Two-dimensional honeycomb network through sequence-controlled self-assembly of oligopeptides. Nat Commun 7:10335
39. Klappenberger F (2014) Two-dimensional functional molecular nanoarchitectures—complementary investigations with scanning tunneling microscopy and X-ray spectroscopy. Prog Surf Sci 89:1–55
40. Parkin G (2006) Valence, oxidation number, and formal charge: three related but fundamentally different concepts. J Chem Educ 83:791
41. Pauling L (1960) The nature of the chemical bond and the structure of molecules and crystals: an introduction to modern structural chemistry. Cornell university press
42. Mond L, Langer C, Quincke F (1890) L.—action of carbon monoxide on nickel. J Chem Soc Trans 57:749–753
43. Nikolaev P, Bronikowski MJ, Bradley R, Rohmund F, Colbert DT, Smith K, Smalley RE (1999) Gas-phase catalytic growth of single-walled carbon nanotubes from carbon monoxide. Chem Phys Lett 313:91–97
44. Kano E, Takeguchi M, Fujita J-I, Hashimoto A (2014) Direct observation of ptterminating carbyne on graphene. Carbon 80:382–386
45. Casari C, Tommasini M, Tykwinski R, Milani A (2016) Carbon-atom wires: 1-D systems with tunable properties. Nanoscale 8:4414–4435
46. Hirsch A (2010) The era of carbon allotropes. Nat Mater 9:868–871
47. Krüger A (2010) Carbon—element of many faces. Wiley-VCH Verlag GmbH & Co. KGaA
48. Geim AK, Novoselov KS (2007) The rise of graphene. Nat Mater 6:183–191
49. Baughman RH, Eckhardt H, Kertesz M (1987) Structure-property predictions for new planar forms of carbon: layered phases containing sp^2 and sp atoms. J Chem Phys 87:6687–6699
50. Hoffmann R (1987) How chemistry and physics meet in the solid state. Angew Chem Int Ed 26:846–878
51. Cretu O, Botello-Mendez AR, Janowska I, Pham-Huu C, Charlier J-C, Banhart F (2013) Electrical transport measured in atomic carbon chains. Nano Lett 13:3487–3493
52. La Torre A, Botello-Mendez A, Baaziz W, Charlier J-C, Banhart F (2015) Strain-induced metal-semiconductor transition observed in atomic carbon chains. Nat Commun 6
53. Szafert S, Gladysz J (2003) Carbon in one dimension: structural analysis of the higher conjugated polyynes. Chem Rev 103:4175–4206
54. Nishide D, Dohi H, Wakabayashi T, Nishibori E, Aoyagi S, Ishida M, Kikuchi S, Kitaura R, Sugai T, Sakata M, Shinohara H (2006) Single-wall carbon nanotubes encaging linear chain C10H2 polyyne molecules inside. Chem Phys Lett 428:356–360
55. Tsuji M, Tsuji T, Kuboyama S, Yoon S-H, Korai Y, Tsujimoto T, Kubo K, Mori A, Mochida I (2002) Formation of hydrogen-capped polyynes by laser ablation of graphite particles suspended in solution. Chem Phys Lett 355:101–108
56. Jin C, Lan H, Peng L, Suenaga K, Iijima S (2009) Deriving carbon atomic chains from graphene. Phys Rev Lett 102:205501
57. Siemsen P, Livingston RC, Diederich F (2000) Acetylenic coupling: a powerful tool in molecular construction. Angew Chem Int Ed 39:2632–2657

58. Klappenberger F, Zhang Y-Q, Björk J, Klyatskaya S, Ruben M, Barth JV (2015) Onsurface synthesis of carbon-based scaffolds and nanomaterials using terminal alkynes. Acc Chem Res 48:2140–2150
59. Moses JE, Moorhouse AD (2007) The growing applications of click chemistry. Chem Soc Rev 36:1249–1262
60. Lautens M, Klute W, Tam W (1996) Transition metal-mediated cycloaddition reactions. Chem Rev 96:49–92
61. Sonogashira K, Tohda Y, Hagihara N (1975) A convenient synthesis of acetylenes: catalytic substitutions of acetylenic hydrogen with bromoalkenes, iodoarenes and bromopyridines. Tetrahedron Lett 16:4467–4470
62. Li P, Wang L (2006) A novel silver iodide catalyzed Sonogashira coupling reaction. Synlett 14:2261–2265
63. Takahashi S, Kariya M, Yatake T, Sonogashira K, Hagihara N (1978) Studies of polyyne polymers containing transition metals in the main chain. 2. Synthesis of poly [trans-bis(tri-n-butylphosphine) platinum 1,4-butadiynediyl] and evidence of a rodlike structure. Macromolecules 11:1063–1066
64. Wong W-Y (2005) Recent advances in luminescent transition metal polyyne polymers. J Inorg Organomet Polym Mater 15:197–219
65. Wong WY, Ho CL (2010) Organometallic photovoltaics: a new and versatile approach for harvesting solar energy using conjugated polymetallaynes. Acc Chem Res 43:1246–1256
66. Green KA, Cifuentes MP, Samoc M, Humphrey MG (2011) Metal alkynyl complexes as switchable NLO systems. Coord Chem Rev 255:2530–2541
67. Harriman A, Ziessel R (1998) Building photoactive molecular-scale wires. Coord Chem Rev 171:331–339
68. Ren T (2005) Diruthenium s-alkynyl compounds: a new class of conjugated organometallics. Organometallics 24:4854–4870
69. Guo S, Kandel SA (2010) Scanning tunneling microscopy of mixed valence dinuclear organometallic cations and counterions on Au(111). J Phys Chem Lett 1:420–424
70. Yam VW-W (2002) Molecular design of transition metal alkynyl complexes as building blocks for luminescent metal-based materials: structural and photophysical aspects. Acc Chem Res 35:555–563
71. Long N, Wong W-T (2014) The chemistry of molecular imaging. Wiley, Hoboken, NJ
72. Whittell GR, Hager MD, Schubert US, Manners I (2011) Functional soft materials from metallopolymers and metallosupramolecular polymers. Nat Mater 10:176–188
73. Lang H, George DSA, Rheinwald G (2000) Bis(alkynyl) transition metal complexes, $R^1C{\equiv}C$–[M]–$C{\equiv}CR^2$, as organometallic chelating ligands; formation of m, h1(2)-alkynylbridged binuclear and oligonuclear complexes. Coord Chem Rev 206:101–197
74. Weber PB, Hellwig R, Paintner T, Lattelais M, Paszkiewicz M, Casado Aguilar P, Deimel PS, Guo Y, Zhang Y-Q, Allegretti F (2016) Surface-guided formation of an organocobalt complex. Angew Chem Int Ed 55:5754–5759
75. (2011) Dewar-chatt-duncanson bonding model. Wiley
76. Yam VW-W, Wong KM-C (2005) Luminescent molecular rods-transition-metal Alkynyl complexes. Springer, pp 1–32
77. Bieri M, Blankenburg S, Kivala M, Pignedoli CA, Ruffieux P, Müllen K, Fasel R (2011) Surface-supported 2D heterotriangulene polymers. Chem Commun 47:10239–10241
78. Lafferentz L, Eberhardt V, Dri C, Africh C, Comelli G, Esch F, Hecht S, Grill L (2012) Controlling on-surface polymerization by hierarchical and substrate-directed growth. Nat Chem 4:215–220
79. Zwaneveld NAA, Pawlak R, Abel M, Catalin D, Gigmes D, Bertin D, Porte L (2008) Organized formation of 2D extended covalent organic frameworks at surfaces. J Am Chem Soc 130:6678–6679
80. Weigelt S, Busse C, Bombis C, Knudsen M, Gothelf K, Lægsgaard E, Besenbacher F, Linderoth T (2008) Surface synthesis of 2D branched polymer nanostructures. Angew Chem Int Ed 47:4406–4410

81. Marele AC, Mas-Balleste R, Terracciano L, Rodriguez-Fernandez J, Berlanga I, Alexandre SS, Otero R, Gallego JM, Zamora F, Gomez-Rodriguez JM (2012) Formation of a surface covalent organic framework based on polyester condensation. Chem Commun 48:6779–6781
82. Dienstmaier JF, Medina DD, Dogru M, Knochel P, Bein T, Heckl WM, Lackinger M (2012) Isoreticular two-dimensional covalent organic frameworks synthesized by on-surface condensation of diboronic acids. ACS Nano 6:7234–7242
83. Abel M, Clair S, Ourdjini O, Mossoyan M, Porte L (2011) Single layer of polymeric Fe-phthalocyanine: an organometallic sheet on metal and thin insulating film. J Am Chem Soc 133:1203–1205
84. Yang B, Björk J, Lin H, Zhang X, Zhang H, Li Y, Fan J, Li Q, Chi L (2015) Synthesis of surface covalent organic frameworks via dimerization and cyclotrimerization of acetyls. J Am Chem Soc 137:4904–4907
85. Perepichka DF, Rosei F (2009) Extending polymer conjugation into the second dimension. Science 323:216–217
86. Wan S, Guo J, Kim J, Ihee H, Jiang D (2008) A belt-shaped, blue luminescent, and semiconducting covalent organic framework. Angew Chem Int Ed 47:8826–8830
87. Grill L, Dyer M, Lafferentz L, Persson M, Peters MV, Hecht S (2007) Nanoarchitectures by covalent assembly of molecular building blocks. Nat Nanotechnol 2:687–691
88. Côté AP, Benin AI, Ockwig NW, O'Keeffe M, Matzger AJ, Yaghi OM (2005) Porous, crystalline, covalent organic frameworks. Science 310:1166–1170
89. Candelaria SL, Shao Y, Zhou W, Li X, Xiao J, Zhang J-G, Wang Y, Liu J, Li J, Cao G (2012) Nanostructured carbon for energy storage and conversion. Nano Energy 1:195–220
90. Lipton-Duffin JA, Ivasenko O, Perepichka DF, Rosei F (2009) Synthesis of polyphenylene molecular wires by surface-confined polymerization. Small 5:592–597
91. Chen Y-C, de Oteyza DG, Pedramrazi Z, Chen C, Fischer FR, Crommie MF (2013) Tuning the band gap of graphene nanoribbons synthesized from molecular precursors. ACS Nano 7:6123–6128
92. Talirz L, Ruffieux P, Fasel R (2016) On-surface synthesis of atomically precise grapheme nanoribbons. Adv Mater
93. Haq S, Hanke F, Dyer MS, Persson M, Iavicoli P, Amabilino DB, Raval R (2011) Clean coupling of unfunctionalized porphyrins at surfaces to give highly oriented organometallic oligomers. J Am Chem Soc 133:12031–12039
94. Zhong D, Franke J-H, Podiyanachari SK, Blömker T, Zhang H, Kehr G, Erker G, Fuchs H, Chi L (2011) Linear alkane polymerization on a gold surface. Science 334:213–216
95. Liu J, Chen Q, Xiao L, Shang J, Zhou X, Zhang Y, Wang Y, Shao X, Li J, Chen W, Xu GQ, Tang H, Zhao D, Wu K (2015) Lattice-directed formation of covalent and organometallic molecular wires by terminal alkynes on Ag surfaces. ACS Nano 9:6305–6314
96. Sun Q, Cai L, Ma H, Yuan C, Xu W (2016) Dehalogenative homocoupling of terminal alkynyl bromides on Au(111): incorporation of acetylenic scaffolding into surface nanostructures. ACS Nano 10:7023–7030
97. Zhang Y-Q, Kepčija N, Kleinschrodt M, Diller K, Fischer S, Papageorgiou AC, Allegretti F, Björk J, Klyatskaya S, Klappenberger F, Ruben M, Barth JV (2012) Homocoupling of terminal alkynes on a noble metal surface. Nat Commun 3:1286
98. Cirera B, Zhang Y-Q, Björk J, Klyatskaya S, Chen Z, Ruben M, Barth JV, Klappenberger F (2014) Synthesis of extended graphdiyne wires by vicinal surface templating. Nano Lett 14:1891–1897
99. Ivanovskii A (2013) Graphynes and graphdyines. Prog Solid State Chem 41:1–19
100. Peng Q, Dearden AK, Crean J, Han L, Liu S, Wen X, De S (2014) New materials graphyne, graphdiyne, graphone, and graphane: review of properties, synthesis, and application in nanotechnology. Nanotechnol Sci Appl 7:1–29
101. Li Y, Xu L, Liu H, Li Y (2014) Graphdiyne and graphyne: from theoretical predictions to practical construction. Chem Soc Rev 43:2572–2586

102. Liu J, Ruffieux P, Feng X, Müllen K, Fasel R (2014) Cyclotrimerization of arylalkynes on Au(111). Chem Commun 50:11200–11203
103. Sun Q, Zhang C, Li Z, Kong H, Tan Q, Hu A, Xu W (2013) On-surface formation of one-dimensional polyphenylene through Bergman cyclization. J Am Chem Soc 135:8448–8451

Chapter 2
Experimental Methods

2.1 Scanning Tunneling Microscopy

The presented results in this thesis are mainly based on STM, which was utilized to analyze the structural and electronic properties of organic molecules adsorbed on metallic surfaces. The purpose of this chapter is to introduce the reader to the basic theory of STM and experimental concepts of the employed STM setup. In addition to STM, X-ray spectroscopy (XS) methods performed by coworkers are introduced as complementary techniques at the end of this chapter.

2.1.1 Basic Principles

For the non-destructive analysis of metal surfaces, Young developed a microscope with a spatial resolution of ∼4000 Å and proposed the idea to use quantum-mechanical tunneling between probe and sample to enhance the instrument's precision [1, 2]. Binnig and Rohrer received the Nobel award in 1986 for realizing the STM as a powerful surface science technique to characterize conducting surfaces with atomic resolution. Its functionality is based on the quantum-mechanical tunnelling current between a sharp metallic tip ideally terminated by one atom and closely approached to a metallic surface. After sufficiently reducing the width of the vacuum barrier by approaching the tip toward the sample (tip-sample distance $z < 1$ nm), electrons are able to tunnel through the vacuum barrier between sample and tip. By applying voltages to piezoelectric elements, the STM tip can be moved with sub-nanometer precision parallel to a conducting sample surface (xy-plane) and perpendicular to it (height z) (see Fig. 2.1b). Thereby, microscopic regions of the sample can be imaged with atomic resolution.

The quantum-mechanical phenomenon of electron tunneling contrasts the classical point of view that a free electron cannot overcome a potential barrier which is higher than its kinetic energy. Figure 2.1a exemplifies electron tunneling between

R. Hellwig, *Alkyne-Based Nanostructures on Silver Substrates*, Springer Theses,
https://doi.org/10.1007/978-3-030-00997-7_2

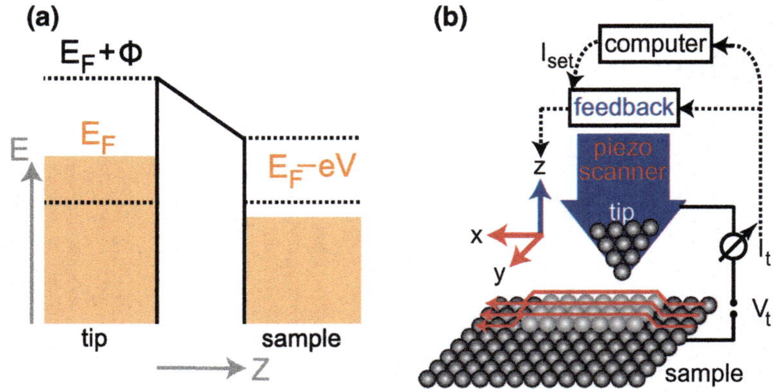

Fig. 2.1 **a** Description of 1D tunneling: Electrons tunnel from occupied tip to unoccupied sample states within an energy window of $E_F - eV \leq E \leq E_F$. **b** Schematic setup of STM. Piezo elements move the tip parallel to the surface (xy-plane) in such a way that the surface topography is imaged in a line-by-line fashion. Within the constant current mode, the tunneling current I between tip and sample is compared to an adjusted setpoint value I_{set} and the feedback circuit corrects the tip height (z-piezo) in such a way that I equals I_{set} during scanning

two metallic electrodes separated by a thin vacuum gap, whereby the energy to extract one electron is represented by the work function Φ. In quantum-mechanics, the free electron is described by a wave, whose amplitude exponentially decays into the barrier region. Accordingly, the square value of the wave function corresponds to a finite probability to encounter the electron at the opposite electrode. Solving the time-independent Schrödinger equation within the semi-classical Wentzel–Kramers–Brillouin (WKB) approximation for 1D electron tunneling through a barrier with varying height yields the transmission coefficient $T(E, z)$ for electron transfer

$$T(E, z) \propto exp(-\frac{2}{\hbar}\int_0^z \sqrt{2m(\Phi(z) - E)}dz, \qquad (2.1)$$

whereby $\hbar$ is Planck's constant, m the mass and E the energy of the electron [3]. For each electrode in Fig. 2.1a, the Fermi-level E_F separates occupied states below E_F from unoccupied states above E_F. The wavefunctions of electronic tip and sample states overlap within the classically forbidden region. If both metals exhibit the same Fermi-level (${E_F}^s = {E_F}^t$), the tunneling rate from sample to tip equals the reverse rate from tip to sample, thus no net current arises. However, when applying a voltage between both electrodes, the Fermi-level of tip and sample shift relative to each other by the energy eV, whereby tunneling from occupied to unoccupied states leads to a finite tunneling current [4]. The summation of all relevant tunneling events within the energy window of width eV leads to the expression

$$I \propto \sum_{E_s=E_F-eV}^{E_F} |\Psi_s(0)|^2 e^{-2\kappa z} \tag{2.2}$$

for the tunneling current, whose magnitude depends on the tip-sample separation z, sample (s) wave function $\Psi_s(z)$, electron energy E_s with respect to E_F, electron probability density $|\Psi_s(0)|^2$ at the sample and the inverse decay constant $\kappa = \sqrt{2m(\Phi - E)}/\hbar$. The current exponentially decays by one order of magnitude, when increasing the distance z by 0.1 nm [5, 6].

Bardeen Formalism

On the basis of time-dependent perturbation theory, Bardeen developed a quantitative approximation for the current passing through a metal-vacuum-metal junction [7]. His model assumes a weak tip-sample interaction treated as small perturbation to the electrodes' eigenfunctions. According to Fermi's golden rule, the current is obtained by summing over all elastic electron transfer processes from filled to empty electronic states within the Fermi distribution $f(E)$ at finite temperatures [5]:

$$I = \frac{2\pi e}{\hbar} \sum_s \sum_t f(E_s)(1 - f(E_t + eV))|M_{st}|^2 \delta(E_s - E_t), \tag{2.3}$$

whereby Ψ_s (Ψ_t) represents the wavefunction of an electronic sample (tip) state with energy E_s (E_t) and M_{st} the associated tunneling matrix element coupling sample and tip states. Upon assuming non-interacting tip and sample eigenfunctions $\Psi(z) = \Psi(0)e^{-\kappa z}$ for both electrodes [8], M_{st} is derived from its overlap at an arbitrary surface between tip and sample (Fig. 2.2a):

$$M_{st} = \frac{\hbar^2}{2m} \int d\vec{S} \cdot (\Psi_t^* \vec{\nabla} \Psi_s - \Psi_s \vec{\nabla} \Psi_t^*). \tag{2.4}$$

The evaluation of this surface integral shows that its square value is identical to the transmission factor $T(E, z)$ in Eq. 2.1 [9]:

$$|M_{st}|^2 = T(E, V, z) \propto e^{-2z\kappa}. \tag{2.5}$$

Assuming that the electrodes are kept at low temperatures, which is true for low-temperature STM (LT-STM) experiments, the Fermi distribution can be approximated by a step function, thus reducing the Eq. 2.3 to [10, 11]:

$$I = \frac{2\pi}{\hbar} e^2 V \sum_s \sum_t T(E, V, z) \delta(E_s - E_F) \delta(E_t - E_F). \tag{2.6}$$

For small voltages compared to the work function, it is found that the current represents a convolution of the density of the states (DOS) of the sample (ρ_s) and the DOS of the tip (ρ_t) [8]:

$$I = \frac{4\pi e}{\hbar} \int_0^{eV} \rho_s(E)\rho_t(E - eV)T(E, V, z)dE. \tag{2.7}$$

To resolve features of the sample DOS at specific voltages and tip locations, a metallic tip with a nearly constant DOS is necessary. Also the strong influence of the transmission factor on the tunneling current needs to be considered. The schematic picture in Fig. 2.2b illustrates tunneling processes arising from a featureless STM tip with flat DOS $\rho_t(E)$, whereby the difference between the vacuum level E_{vac} and E_F corresponds to the work function Φ. The colored region and horizontal arrows in Fig. 2.2b, c indicate elastic transitions within the energy interval eV. When applying a positive voltage on the sample, the current flows from the tip to the sample, while a negative voltage initiates electron tunneling from occupied sample states to unoccupied tip states ($E_F{}^s = E_F{}^t + eV$). As exemplarily shown by Feenstra et al. for a GaAs(100) surface, voltage-dependent STM imaging can give information on chemical aspects. They could assign the respective lattices of Ga and As atoms by inspecting the spatial distribution of occupied (unoccupied) DOS within negative (positive)-voltage STM images [9, 12].

Interpreting the matrix elements in Eq. 2.5 for small voltages at the energy E_F, Tersoff and Haman extended the 1D Bardeen approach by additionally taking into account the three dimensional (3D) shape of the STM tip. Upon the simplified assumption of an isotropic 3D tip apex exhibiting s-wave orbital character, the current reads

$$I \propto \sum_s |\Psi_s(\vec{r}_0)|^2 \delta(E_s - E_F) \equiv \rho_s(E_F, \vec{r}_0), \tag{2.8}$$

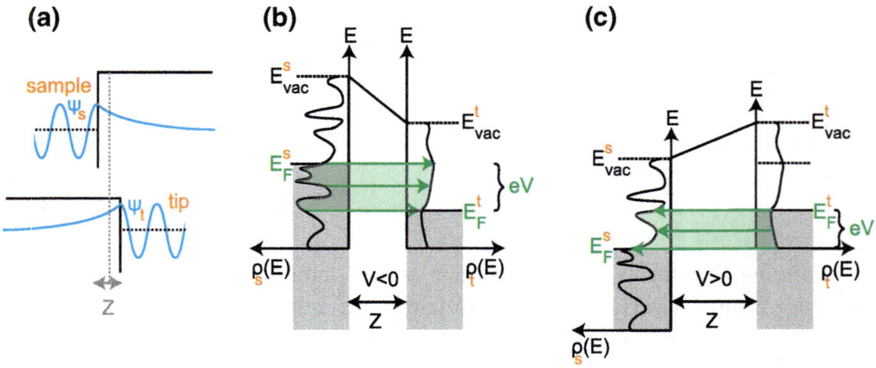

Fig. 2.2 **a** Bardeen approach. The tip-sample interaction is evaluated from the electrodes' eigenfunctions. **b** Summing up all tip-sample transitions from occupied to onuccupied states results in the total tunneling current. **c** Reverse current flow upon reversal of voltage polarity. Indices s and t refer to sample and tip

whereby $|\Psi_s(\vec{r}_0)|^2$ represents the electron probability density at the STM tip. Within this approximation, the current is related to the local density of states (LDOS) of the sample's Fermi contour at the tip position $\vec{r}_0$ [9, 13]. Albeit the s-wave model does not describe a real tip and tip-sample interactions are not negligible at sufficiently high voltages and small tip-sample distances, STM simulations based on Tersoff-Hamann theory often correspond well with the experimental findings. Further refinements of this model include wave-vector dependent tunneling (Baratoff et al. [14]) and the inclusion of tip wave functions with higher orbital numbers (Chen et al. [13]) [8].

Modes of operation

There are two modes for obtaining a topographic image of a surface region (Fig. 2.3a, b): In the **constant height mode**, the tip height remains constant, while the current $I(x, y)$ is recorded at each sample position. This mode allows high scanning speed but is restricted to atomically flat surface regions. It is often applied for acquiring spectroscopic conductivity maps, for which z-corrections of the feedback loop are not desired.

The usual scanning operation is the **constant current mode**. This mode is based on a feedback loop mechanism that corrects the tip height until the current equals the adjusted setpoint value I_{set} (cf. Fig. 2.1b). While the current signal remains constant, the recorded tip height $z(x, y)$ is a function of the sample position and describes the surface topography. Within the approximations mentioned above, a topographic constant-current 2D map can be related to the Fermi-level DOS contour of the sample surface.

Scanning Tunneling Spectroscopy

Besides spatial STM imaging, voltage-current curves at well-defined locations reflect the sample LDOS with uttermost spatial and energy resolution. Already before the invention of STM, $I(V)$ spectroscopy on tunneling junctions performed by Giaever [15] showed the linear increase of the current with the voltage, whilst the DOS of both electrodes remains constant [15]. Nowadays, the local character of low-temperature STS allows the LDOS analysis of individual surface atoms and adsorbates. Due to the fact that the current in Eq. 2.7 corresponds to an integral over the energy window eV, the spectroscopic information is better resolved by its first derivative. Within the

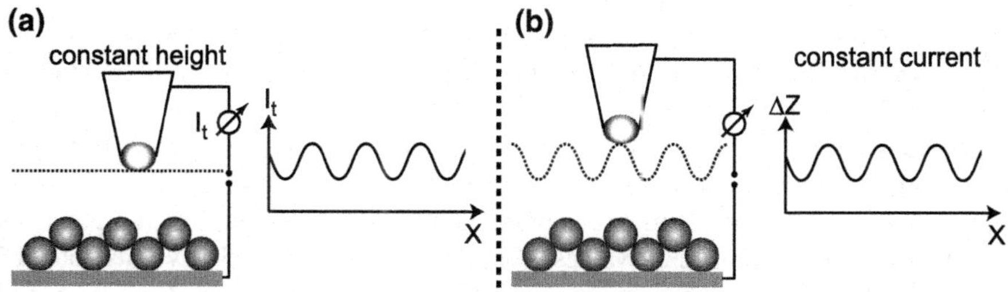

Fig. 2.3 STM imaging modes: at constant height (**a**) or constant current (**b**)

most simple model assuming constant ρ_t for the tip, small voltages and neglectable variations of $T(E, V, z)$, the differential conductivity at the voltage V can be related to the LDOS of the sample (ρ_s) at the energy eV:

$$dI/dV \propto \rho_s(eV). \tag{2.9}$$

However, one must be careful in interpreting spectral features at high voltages or comparing spectra recorded at different tip-sample distances $\Delta z(x, y)$. Under these conditions, STS spectra can show significant deviations of the spectral background signal due to voltage- and z-dependent variations of $T(E, V, z)$. For this reason, the normalization $\frac{dI/dV}{I/V}$ is often used to cancel the detrimental effects of the transmission factor [9, 16].

To obtain a dI/dV spectrum with a high signal-to-noise ratio, the lock-in technique is used instead of numerical differentiation of the $I(V)$ spectrum. The advantage of integrating an external lock-in amplifier (LIA) into the STM electronics is based on its selectivity toward detecting and enhancing a single signal of defined frequency and phase from a noisy background. The LIA technique requires a sinusoidal modulation voltage V_{mod} of certain frequency ω and phase to be added to the tunneling voltage V_{t}. As a consequence, the current expresses an oscillating component I_{mod} according to the expression

$$I(t) = I(V_{\text{t}} + V_{\text{mod}} sin(\omega t)) = I_{\text{t}} + I_{\text{mod}}. \tag{2.10}$$

A Taylor expansion of I_{mod} in terms of V_{mod} yields

$$I_{\text{mod}} \propto \underbrace{\frac{dI(V)}{dV}}_{\propto \rho_S(eV)} V_{\text{mod}} cos(\omega t) + \underbrace{\frac{d^2I(V)}{dV^2}}_{IETS} \frac{V_{\text{mod}}^2}{4} cos(2\omega t). \tag{2.11}$$

This implies that the first harmonic signal detected by the LIA at ω represents the first derivative of the current, which is related to ρ_s. To obtain a STS spectrum, the open feedback loop (OFL) mode is used while ramping V_{t} and forwarding I_{t} to the LIA (Fig. 2.4). Its output signal used to build up the $dI/dV(V)$ spectrum is proportional to the change in current in the vicinity of V_{t} [17]. As the current is composed of elastic and inelastic tunneling processes, second harmonic detection is interesting for inelastic tunneling spectroscopy (IETS) experiments. By acquiring $\frac{d^2I(V)}{dV^2}$ spectra, the excitation of vibrational and rotational degrees of freedom of single molecules [18, 19] or spin-flip excitations [20] can be studied.

Besides taking single spectra, the spatial localization of a spectral DOS feature can be imaged by taking a dI/dV map in the OFL mode [21]. Within the performed experiments, typical values for the applied modulation amplitude V_{mod} ranged from 1 mV rms (root-mean-square values) for low-bias spectra to 5 mV rms for STS on molecular orbitals. The applied oscillation frequencies were $f = 733.5$ and 933.5 Hz.

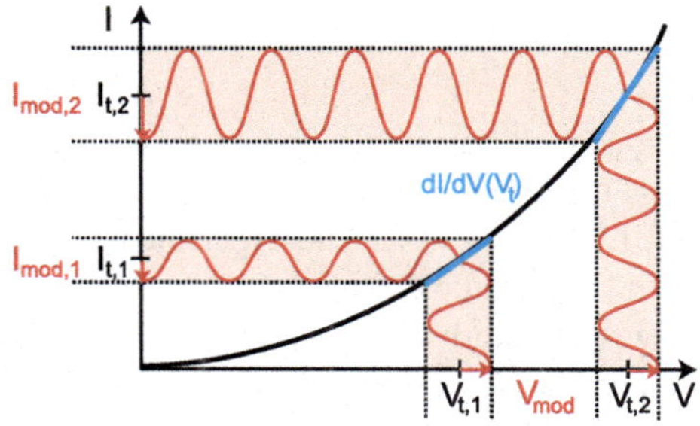

Fig. 2.4 Lock-in technique. A small sinusoidal signal added to the bias modulates the tunneling current

STM Tip Manipulation

The atom-precise positioning of the STM tip is ideally suited for the manipulation of single adsorbates. The uptake of single atoms or molecules can be used for changing the chemical nature or imaging contrast of the tip, changing the adsorbate structure (bond scission), or bottom-up creation of an artificial nanostructure. There are two kinds of manipulation experiments.

Vertical Manipulation

By vertically approaching the tip placed above an adsorbate toward the surface, single atoms or molecules can be picked up by forming chemical bonds between tip apex and adsorbate (Fig. 2.5a). The adsorbate attached to the tip apex can be dragged to a new location and dropped in a reversible fashion.

Lateral Manipulation

In the first step, vertical manipulation is used to establish a bond between tip apex and adsorbate. The STM tip is approached next to the adsorbate (or on top of it) before guiding it across the surface to the desired location (Fig. 2.5b). Meanwhile, adsorbate-tip interactions lead to a pushing, sliding, or pulling mode influenced by the substrate [22]. The tip-guided displacement of compounds and oligomers can indicate the presence of on-surface formed bonds.

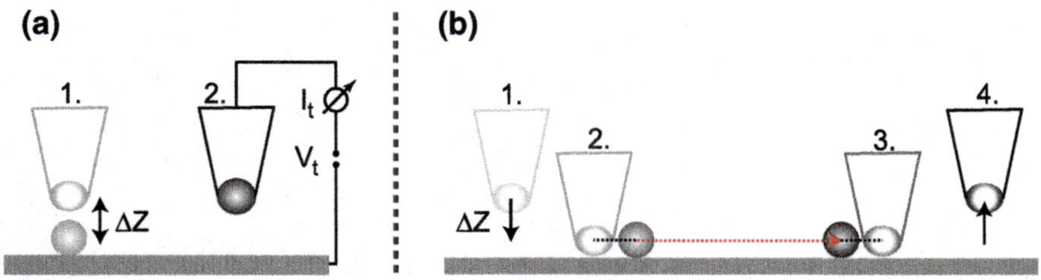

Fig. 2.5 Schematic drawing depicting vertical manipulation (**a**) and lateral manipulation (**b**) of an adsorbate via the STM tip

2.1.2 Experimental Setup

All measurements were carried out with a commercial Joule-Thompson Scanning Tunneling Microscope (JT-STM) [24] operating under UHV conditions and at variable temperatures from 1.2 up to 400 K. The experimental setup consists of preparation and STM chamber separated by a gate valve, as shown in Fig. 2.6. The former is utilized for preparing a sample by the standard techniques for sample cleaning and preparation, while the second chamber accommodates the STM instrument for sample characterization. For the contamination-free prepararation and diffusion-free characterization of surface-adsorbed nanostructures, UHV and cryogenic conditions are mandatory. Therefore, the STM is connected to a triple-stage cryostat which reaches a continuous sample temperature down to 1.2 K. Since STM measurements are very susceptible to mechanical vibrations, the experimental setup is carried by four pneumatic damping feet decoupling it from low-frequency noise.

Ultra-High Vacuum

When splitting a single crystal in an environment of 10^{-6} mbar, the time left to characterize a well-defined substrate would be limited to split seconds, during which the surface gets rapidly covered and contaminated by adsorbing residual gas atoms. Therefore, UHV pressures in the 10^{-10} mbar regime are required to maintain the prepared sample sufficiently clean during several days [25]. Once the sample is tansferred to the STM and the tip approached for tunneling, the cryogenic environment of the JT-STM allows long-term characterization with lateral drifts below 20 pm per hour [26].

To introduce a new STM tip or crystal from ambient to UHV environment, it is first of all attached to the transfer bar of the load-lock, which is pumped by a

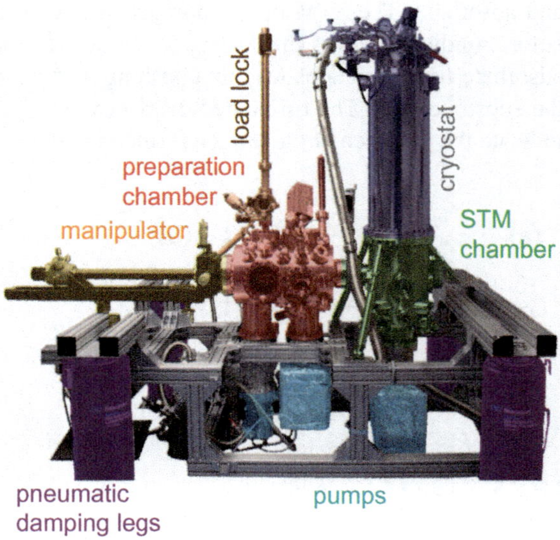

Fig. 2.6 Picture [23] of the JT-STM setup consisting of preparation and STM chamber connected to cryostat. The pumping system maintains ultra-high vacuum conditions

pumping station. After, it is attached to the manipulator equipped with a heating oven, cooling station and temperature readout. The UHV system of the chamber consists of a triple-stage pumping system monitored by cold cathode gauges [27]. The first stage is a diaphragm pump [28] providing a prepressure of 3.5 mbar for a small turbo pump [29] reaching $5 \cdot 10^{-9}$ mbar. The third stage is a turbomolecular pump with magnetic bearing [30], which reaches UHV pressures. To quench mechanical noise during STM operation, the mechanical pumps can be switched off, while ion getter pumps maintain UHV conditions. The pressure can be further optimized via a titanium sublimation pump based on the sorption of residual gas atoms at reactive Ti sites during the resistive evaporation of Ti from a filament [25]. With an in situ installed residual gas analyzer [31], the chemical purity of the material from the evaporator can be checked during heating. Once a successful sample preparation is finished, the manipulator head with sample is precooled by liquid nitrogen (l N_2) while the sample transfer to the STM is prepared.

Joule-Thompson Cryostat

Figure 2.7a shows the three cooling stages of the Joule-Thompson (JT) cryostat. The outer reservoir of l N_2 (77 K) minimizes the consumption of liquid He (l He) within the inner He tank (4 K). For the activation of the JT cooling stage, He gas is filled from an outside gas cylinder through a long pipe, precooled from room temperature (RT) to liquid He temperatures on the way down to the JT pot (Fig. 2.7c). The liquefying He gas fills the JT pot, thus leading to a slight drop of the STM temperature. After, the valve to the running JT pump is opened in order to pump the He condensate within the JT pot through a pumping line. Thereby, He vapor is formed via Joule-Thompson expansion through a tiny capillary with a diameter of 90 μm [26], which leads to an operation temperature of 1.2 K. The JT pot needs to be thermally decoupled from the cooling plate of the He tank via a wooble stick in order to maintain the low temperature. A further reduction in temperature ($\sim$1 K) is achieved by closing the He inlet to the JT pot, thereby allowing optimal performance until the condensate is consumed (single-shot mode).

Noise Reduction

To optimize the STM and STS performance, different strategies are implemented to decouple and reduce unwanted vibrations, i.e.,, the mechanical and electronical noise frequencies identified through a Fast Fourier Transform (FFT) spectrum of the tunneling current. Firstly, the whole experimental setup is standing on four pneumatic damping feets. Secondly, the STM stage is decoupled from the surroundings via three copper beryllium springs, which significantly reduce the noise level. In addition, copper beryllium springs and damping elements were mounted between the l N_2 and l He tanks to damp resonance frequencies arising from the suspension system of the cryostat and bubbling of cryogenic liquids. Furthermore, an eddy-current mechanism based on fixed magnets opposite to copper wings (Fig. 2.7b) is mounted below the STM stage, in order to reduce oscillations of the STM head. To diminish capacitance effects and crosstalk within cables, special coaxial wires are used for current and voltage lines; feed lines to piezo units and temperature sensors are low-pass filtered [26].

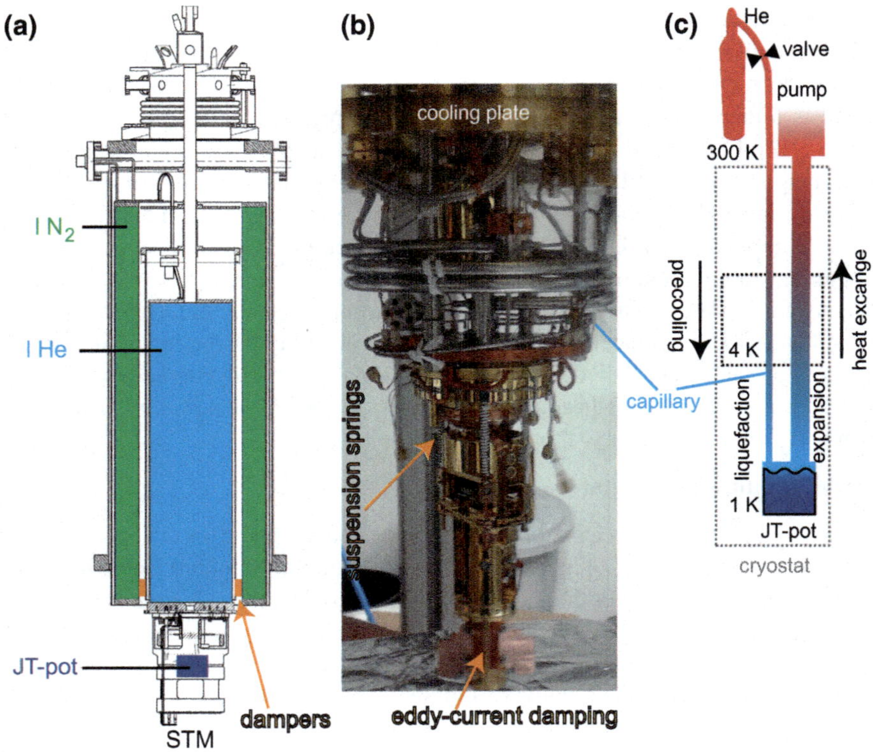

Fig. 2.7 **a** Technical drawing [32] of the Joule-Thompson (JT) cryostat comprised of three cooling stages: l N_2 tank, l He tank and JT stage. **b** JT stage with STM stage hanging on three suspension springs. **c** Schematic overview on the working principle of the JT stage. Precooled He gas liquefies in the JT pot, which is subsequently pumped to initiate JT expansion

The STM Stage

The STM body located inside of three radiation shields below the cryostat tanks (Fig. 2.8a) is made of stiff material with low eigenfrequency and high thermal conductivity. The photo in Fig. 2.8b shows the access to the sample rack holding the sample, i.e., a single crystal mounted on Mo plates (not shown). During sample transfer, a wobblestick [33] is used to pick up the precooled sample from the manipulator and carefully slide it into the sample rack, where it is held by Mo springs. Subsequently, the sample rack is positioned by driving the amplitude and frequency of a sequence of voltage pulses addressed to the Y-motor (range: 5 mm). In detail, six clamped shear-piezo elements facilitate a sliding movement between ruby spheres and saphire platelets [26].

During the coarse approach of the tip, an etched tungsten wire mounted at the STM tip holder is moved upwards by means of six shear-piezos on sliding saphire surfaces (Z-motor range: 10 mm). The prism design of the scanner holder and working principle of the Z-motor was developed by Pan [34]. After the coarse approach, the radiation shields are closed in order to cool down the sample and further approach the

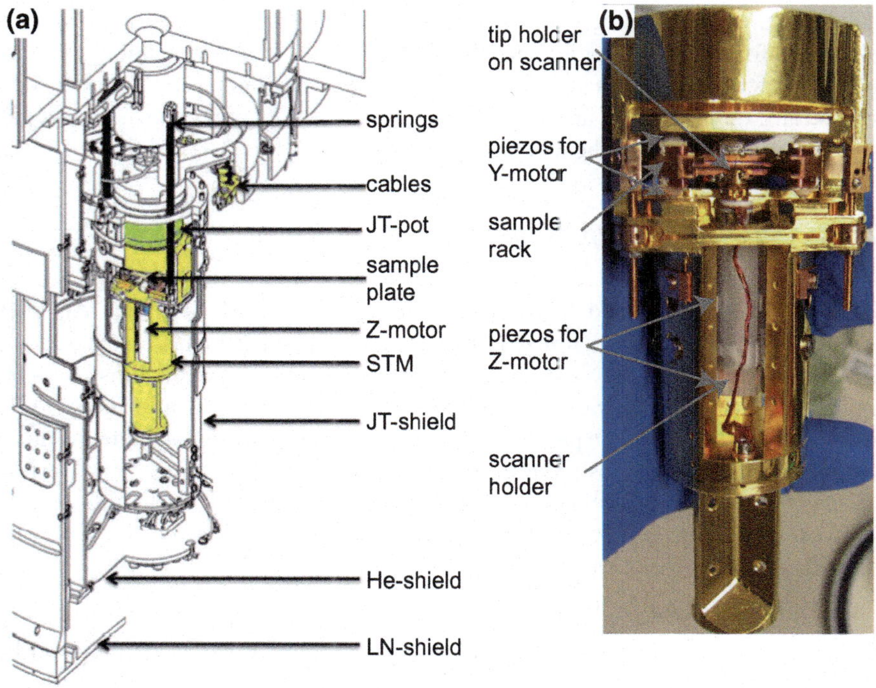

Fig. 2.8 Schematic drawing [24] (**a**) and picture (**b**) of the STM head. The sample rack can be moved in one direction by piezo motors. Coarse motors for lateral coarse movement of sample (Y-motor) and Z-approach of the tip (Z-motor below) are indicated. For scanning, controlled voltage pulses are applied to the piezo segments of the scanner

tip in smaller increments [26]. Two cryogenic temperature sensors [35] located close to the sample and at the JT pot allow precise temperature detection. The massive scanner holder made from Cu and thermally connected to the STM body allows fast thermalization after coarse motion and constant temperature during scanning. The scanner design enables the staightforward exchange of the STM tip holder by hooking it into a special Mo plate and moving tip and plate via the coarse motors [26].

Scanning Piezo

Figure 2.8b shows the scanning piezo tube, which is integrated in the scanner holder and carries the tip holder. At l He temperatures, the piezo elements ensure a XY scanning window of $2 \times 2\ \mu m^2$ and Z-range of 200 nm [26]. When tip and sample are within the tunneling regime, an active feedback loop is used to control the tip-sample distance by correcting the Z-height (constant current mode). By the simultaneous control of different piezo segments, the scanning of a sample region is realized through this mode.

Data Processing

The output signal of the tunneling current is amplified by a preamplifier [36] and forwarded to the Nanonis control unit [37]. Being comprised of a real-time controller,

22 bit multi-channel analog↔digital converter unit, high voltage supply and scanner controller for the piezo motors, the latter represents the signal processing interface betwen experimental setup and Nanonis control software. For dI/dV spectroscopy, we use an external lock-in amplifier [38] connected to a Nanonis signal divider and signal adder unit [39] for the modulation voltage V_{mod} [26].

2.1.3 Sample and Tip Preparation

For the projects presented in this thesis, the single crystal substrates Ag(111), Ag(100) and Ag(110) were utilized [40]. The respective facet characterized by the cutting direction of the crystal defines the arrangement of surface atoms: Ag(111) exhibits a sixfold symmetric and dense-packed atomic arrangement, while Ag(100) shows a fourfold symmetric packing. Ag(110) expresses a twofold-symmetric lattice with topmost surface atoms arranged in dense-packed lines alternating with surface grooves. The symmetry, atomic registry and chemical aspects of each of the chosen substrates can strongly influence the self-assembly and on-surface reaction pathways of surface-adsorbed molecules.

Surface Preparation

In order to remove adsorbates from the sample surface, it has to be prepared by repeated cycles of sputtering and annealing. To this end, Ar gas is introduced in the preparation chamber through a leak valve (pressure of $\sim 3 \cdot 10^{-5}$ mbar). Then, a voltage of 0.9 kV is applied to a sputter gun in order to ionize Ar atoms along the trajectory of a high-energy electron beam directed toward the sample. During the sputtering process (25 minutes (min) duration), a current of Ar ions (corresponding to 10 μA emission current) removes topmost surface material. To obtain a smooth and homogenous surface, the sample is subsequently annealed at 760 K (for 15 min) by resistively heating it via a current-driven tungsten filament (maximal filament setpoint: $V = 12$ V, $I = 2.5$ A). Higher sample temperatures are accomplished by electron impact heating via the use of a voltage gradient of 750 V between sample and filament. Thereby, electrons emitted from the filament deposit their kinetic energy on the backside of the sample. The sample temperature is read from a pair of chromel-alumel wires (Type K) which exclusively contacts the crystal edge. The sputtering and annealing cycle were repeated until STM imaging reveals a clean and smooth surface.

Sample Preparation

The synthesis of low-dimensional nanostructures was accomplished through the deposition of organic molecules and metal atoms on clean metal surfaces. All deposition steps were performed in the preparation chamber, except the in situ deposition of Co atoms.

Customized Molecules

The molecules were evaporated on clean substrates (organic molecular beam epitaxy (OMBE)) from a high-purity powder filled in a clean quartz crucible within an evaporation cell of a home-made triple evaporator (see third chapter of Ref. [41] for technical details). The employed alkyne-functionalized molecules were de novo synthesized by the chemists Zhi Chen and Ping Du from the Ruben group [42]. Initially, the synthesized molecular powder was thoroughly degassed under UHV conditions in order to get rid of volatile byproducts. The resistance of the heating wire (platin) wound around the crucible was measured in order to obtain the crucible temperature after each current-heating pulse. The degassing step requires caution since the reactive alkyne group is prone to polymerization processes in the evaporation cell. Prior to sample preparation and STM characterization, the cleanliness of the molecular powder could be further checked by a residual gas analyzer [31]. After reaching stable evapoaration conditions, the sample was positioned in front of the molecular crucible, before opening the shutter for a defined time.

Metal Elements

Transition metal atoms were sublimated from a thin wire (99.995% purity for Co [43]), which was wound around a tungsten wire resistively heated by a direct current. Lanthanide atoms were sublimated from a foil (99.9% purity for Ho [44]), which was clamped by copper rods carrying an alternating current. After thoroughly degassing the high-purity material in UHV and depositing specific amounts of it on the (precooled) sample, the deposition flux and adsorbate concentration could be calibrated via STM imaging and was adjusted through filament heating current and deposition time. For low-temperature deposition, variable sample temperatures down to 100 K are possible through l N_2-cooling of the manipulator whilst counterheating the sample. After the evaporation, the sample can be annealed to enhance molecular diffusion and activate on-surface reactions. For the in situ deposition of single Co atoms, an evaporator placed in front of the cryostat's evaporation gate was opened for a certain time, during which the sample temperature did not exceed 12 K.

Tip Preparation

Atom-precise scanning probe microscopy relies on the chemical purity and shape of the scanning tip apex. Such requirements are fulfilled by a sharp tungsten tip, which is created by the electrochemical etching of a tungsten wire. To this end, a W wire is centered within a ring-shaped counterelectrode (gold) that carries a lamella of NaOH solution (Fig. 2.9a) [45]. Unless the NaOH lamella does not break before, the process terminates with a meniscus-shaped tip falling of the W wire and smoothly landing in a foam-like environment. After the tip is cleaned in destillated water, mounted on the tip holder, introduced into UHV, sputtered from different angles and mounted into the STM, residual amounts of an oxide layer are removed via field emission. Within this procedure (Fig. 2.9b), a high negative voltage is applied to the STM tip while carefully approaching it toward the clean sample until reaching an electron current in the order of μA. The current, limited by a high-value resistance, is due to the field-induced lowering of the tip's work function at nano-scaled tip-sample distances. Afterward, the polarity of the external voltage supply is reversed and the tip is slightly brought into slight contact with a clean sample spot and subsequently withdrawn. Thereby, a clean and sharp tip is created and coated with substrate atoms.

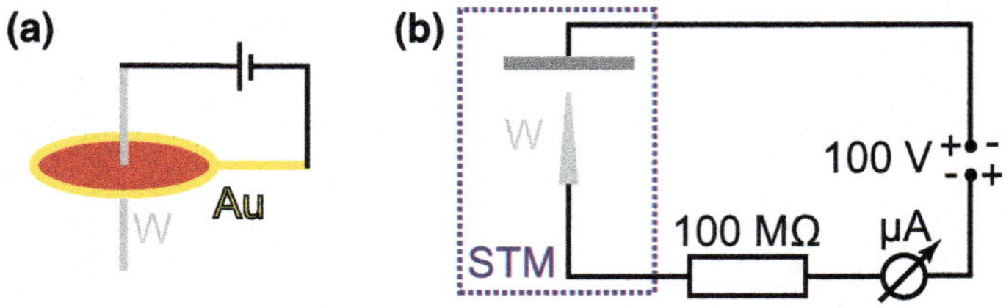

Fig. 2.9 Preparation of the STM tip. **a** Electrochemical etching procedure. **b** Field-emission procedure under UHV conditions

Since the STM signal represented by the tunneling current is a convolution of both the sample's and tip's DOS, metallic tips with flat DOS are especially important for STS. For STS, voltage pulses and tip-sample contacts at pristine spots of a clean metal surface are utilized to obtain a reproducible tip.

2.2 X-ray Spectroscopy Techniques

While the local character and high energy resolution of STM permits topographic and spectroscopic elucidation of single-adsorbate species and their assemblies within small domain regions, its drawback is the limited explanatory power on chemical and conformational features which cannot easily be disentangled from the tunneling signal representing a convolution of electronic and geometric aspects. To obtain information on molecular conformation, oxidation states, chemical states and electronic aspects, complementary X-ray photoelectron spectroscopy (XPS) and near-edge X-ray absorption fine-structure (NEXAFS) measurements were performed by coworkers. These XS techniques are surface-sensitive, element-selective and extend the local picture of LT-STM by reflecting ensemble properties dominating within the macroscopic region of the X-ray beam. Accordingly, the combined XS+STM method is a very powerful methodology to obtain key insights on adsorption structure and physicochemical aspects under UHV conditions [46].

Since a monochromatic X-ray beam is required, XS is ideally carried out with synchrotron light sources offering monochromatized light with high brilliance, high polarization, high intensity and tunable energy. The X-ray absorption profile reveals for each atomic species characteristic core level excitations and absorption edges. The core hole vacancy can be successively filled by an outer-shell electron and emit characteristic X-ray photons or secondary electrons (Auger effect), whose energy distribution reflect the atom's energy levels [47]. While XPS was used to address the chemical changes in surface-molecule properties and bonding aspects, NEXAFS results delineate the geometry and alignment of molecular orbitals within functional groups according to the building block principle [48].

2.2.1 X-ray Photoelectron Spectroscopy

X-ray photoelectron spectroscopy (XPS) is based on the light-assisted generation of free electrons escaping from a solid, reminiscent of the photoelectric effect. For monochromatic X-ray photons, photoabsorption sets in when the energy suffices to lift an electron to an unoccupied orbital, or the continuum. The latter case leads to primary electrons of defined kinetic energy E_{kin}, which are collected by a spectrometer during an XPS experiment. Their maximal energy is limited by the applied photon energy Φ and work function of the material. Within the one-electron approximation, the energetic distribution of the photoelectrons (intensity versus binding energy (BE)) obeys energy conservation according to the equation

$$E_b = h\nu - E_{kin} - \Phi, \tag{2.12}$$

whereby E_b is the BE of the ionized core level. When using a photon energy that ionizes the deepest lying core level (K shell), also energetically higher lying levels will give rise to a series of related peaks [49]. Counting and sorting the number of electrons according to their BE, an intensity-versus-BE spectrum of the emitted electrons reflects the chemical composition of the sample. Each atomic species reveals a set of characteristic photoemission peaks, whose BE serving as a fingerprint for its chemical state depend on its chemical enironment.

After the core ionization, the excited (N-1) electron system relaxes back to its ground state, whereby the electrons rearrange and relaxation energy may be transferred to the photoemitted electron. Because of the recombination of the core hole, secondary processes leading to the escape of Auger electrons or fluorescence photons are present. For the light elements C, N, O that are relevant for many organic molecules, the Auger yield exceeds the fluorescence yield by more than two orders. The escape depth of secondary electrons is found to follow a universal behavior (mean free path versus kinetic electron energy) for all materials with a mean free path of less than 10 Å at the relevant X-ray energies between 250 and 600 eV. Since only electron cascades emanating within a topmost surface layer of 10 Å may escape, the electron yield is extremely surface sensitive [48]. During photoexcitation, the interaction of the core hole with the valence electrons of adjacent atoms strongly influences its BE. Comparing the peak positions with reference literature, one can detect changes in the chemical environment of the pertaining element. The so-called chemical shift is sensitive to the effective charge, electronegativity of neighboring atoms and charge transfer processes [50]. The removal of valence electrons from the excited atom leads to an increase in its effective atomic number for the 1s electron, and hence higher BE of the core electrons [25]. XPS is ideally suited to quantitatively monitor chemical changes of molecular functional groups during on-surface reaction steps triggered by sample annealing, e.g., C1s and O1s spectra revealing the deprotonation of carboxylate groups [51], or O1s and Fe3p profiles evincing metal-organic coordination with Fe atoms [52].

2.2.2 *Near-Edge X-ray Absorption Fine-Structure Spectroscopy*

While XPS experiments quantitatively elucidate the chemical state of surface-adsorbed molecules, the synchtrotron-based technique near-edge X-ray absorption fine-structure (NEXAFS) utilizes monochromatic soft X-rays of high intensity and high degree of polarization to study molecular geometries and conformations. Both XPS and NEXAFS detect electrons, however, the former technique analyzes their energy at constant photon energy, while NEXAFS detects the number of electrons as a function of the photon energy, without resolving electron energies. The power of NEXAFS lies in probing the transitions from the deepest lying core level, e.g., C1s to unoccupied bound or continuum states. It is a very sensitive method capable of discriminating conformational states of functional groups, hybridization states, and provides hints to molecular orbitals engaged in chemisorption bonding.

In a typical NEXAFS spectrum, the photon energy is ramped from slightly below the adsorption edge up to 50 eV above it [46, 53]. The absorption probability as a function of the photon energy is related to transitions from the lowest-lying core level to unoccupied states below and above the vacuum energy (Fig. 2.10). The sharp π^* resonances within NEXAFS curves allow a clear distinction of sp^2- and sp^3-hybridized carbon atoms within different chemical environments [53], while σ^* resonances are identified as broad peak profiles above the vacuum level. Accord-

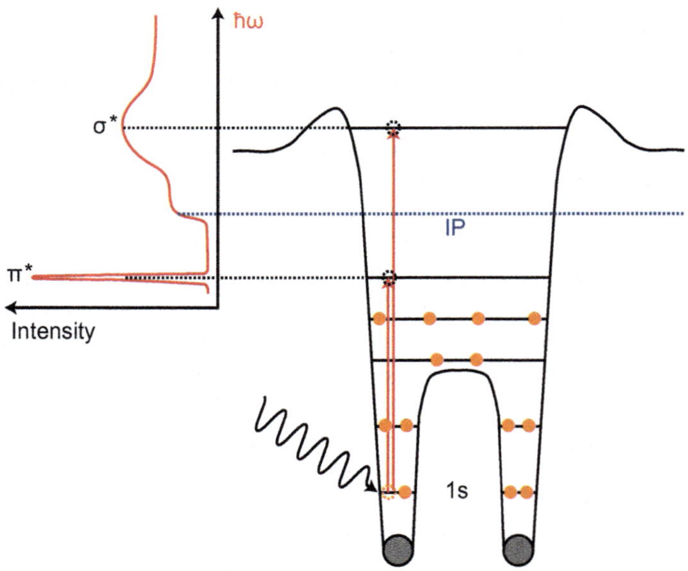

Fig. 2.10 Schematic description of NEXAFS. The potential curve of a diatomic molecule is shown. X-rays eject core electrons probing unoccupied orbitals below and above the ionization threshold. The NEXAFS spectrum shows a sharp π^* and a broad σ^* resonance

ing to the building block principle, large molecules including polymers containing saturated and unsaturated functional moieties are approximated as the superposition of diatomic entities. If the molecular structure contains further atoms besides carbon, complementary information on the assembled molecular layer is obtained by scanning over the respective adsorption edges, e.g., N1s and O1s [54, 55].

Because of the directional aspects of antibonding molecular orbitals, there is a direct relationship between orbial alignment and molecular conformation. Therefore, the intensity I_{if} of peaks related to transitions from the lowest-lying core level (initial state i) to a molecular orbital (final state f) depends on the incident angle of the polarized X-rays, provided that the molecular layer is ordered. For the K-shell excitation of a molecular orbital, the transition intensity

$$I_{if} \propto |\vec{E} \cdot \langle f|\vec{r}|i\rangle|^2 \propto |\vec{E} \cdot \vec{O}|^2 \propto cos^2\delta \tag{2.13}$$

exhibits a strong dependence on the angle δ between the electric field vector $\vec{E}$ and the vector $\vec{O}$ of the largest amplitude of the final state orbital. Comparing NEXAFS spectra for different sample orientations with respect to a fixed direction of the X-ray beam, it follows that the spectral peaks related to σ^* and π^* show opposite angular behavior. The comparison and fit of NEXAFS resonances within spectras recorded for grazing, normal and magic angle incidence yields bonding angles, hence allowing a reconstruction of the bonding geometry of the adsorbed molecules [54].

To obtain a maximal signal-to-noise sensitivity, photoelectron emission related to photoabsorption within the bulk needs to be minimized. This can be achieved by increasing the surface sensitivity via electron detection in the partial mode, where retarding voltages are used to only detect "fast" electrons (from the topmost layers) [53]. For each spectrum shown in this work, a background spectrum obtained from a clean substrate was substracted. Subsequently, the difference spectrum was normalized by its associated transmission spectrum, which was simultaneously obtained from a Au grid transmitted by X-rays. This correction procedure is necessary to compensate artifacts originating from carbon impurities within the beamline optics and instabilities of the photon beam.

2.2.3 Experimental Details

XPS and NEXAFS experiments were conducted by coworkers at the BESSY II synchrotron (HE-SGM beamlines) in Berlin. The X-rays were monochromatized through a grating of 1500 l/mm and slit widths of 200 μm. To determine the kinetic energy of secondary electrons, the pass energy of the hemispherical analyzer (Scienta R3000) was adjusted to 20 eV for the C1s region. An X-ray photon energy of 435 eV was used to obtain free electrons with $E_{kin} = 150$ eV. The substrate's Ag $3d_{5/2}$ peak at 368.3 eV was utilized to calibrate the BE axis. For evaluating the XPS data, a Shirley background was substracted before fitting each spectral peak by a Voigt profile.

2.3 Simulations

Structure models of the molecules were created via the Chemdraw software [56]. If not stated otherwise, gas-phase models of the employed molecular structures were simulated through the Hyperchem software [57], in the framework of the semi-empirical routine AM1 [58]. Theoretical insights on the on-surface adsorption geometry and lowest-energy structure were achieved through state-of-the-art density functional theory (DFT) modeling performed by coworkers, as mentioned in the respective chapters. The visualizations of the models were rendered via the VMD software [59].

References

1. Young R, Ward J, Scire F (1971) Observation of metal-vacuum-metal tunneling, field emission, and the transition region. Phys Rev Lett 27:922
2. Young R, Ward J, Scire F (1972) The topografiner: an instrument for measuring surface microtopography. Rev Sci Instrum 43:999–1011
3. Nolting W (2002) Basic course theoretical physics 5/1 (quantum mechanics). Springer, Heidelberg
4. Wiesendanger R, Güntherodt H-J (1996) Scanning tunneling microscopy III: theory of STM and related scanning probe methods. Springer Verlag
5. Stroscio JA (1993) Scanning tunneling microscopy. Springer, Heidelberg
6. Chen CT, Idzerda YU, Lin H-J, Smith NV, Meigs G, Chaban E, Ho GH, Pellegrin E, Sette F (1995) Experimental confirmation of the X-ray magnetic circular dichroism sum rules for iron and cobalt. Phys Rev Lett 75:152–155
7. Bardeen J (1961) Tunnelling from a many-particle point of view. Phys Rev Lett 6:57–59
8. Meyer E, Hug HJ, Bennewitz R (2013) Scanning probe microscopy: the lab on a tip. Springer Science & Business Media
9. Voigtlaender B (2015) Scanning probe microscopy: atomic force microscopy and scanning tunneling microscopy. Springer Verlag
10. Chen CJ (1993) Introduction to scanning tunneling microscopy. Oxford University Press
11. Tersoff J, Hamann DR (1985) Theory of the scanning tunneling microscope. Phys Rev B 31:805–813
12. Feenstra RM, Stroscio JA, Tersoff J, Fein AP (1987) Atom-selective imaging of the GaAs(110) surface. Phys Rev Lett 58:1192–1195
13. Chen CJ (1992) Effects of m 6 = 0 tip states in scanning tunneling microscopy: the explanations of corrugation reversal. Phys Rev Lett 69:1656–1659
14. Baratoff A (1984) Theory of scanning tunneling microscopy-methods and approximations. Phys B+C 127:143–150
15. Giaever I (1960) Electron tunneling between two superconductors. Phys Rev Lett 5:464–466
16. Feenstra RM, Stroscio JA, Fein AP (1987) Tunneling spectroscopy of the Si (111) 2 × 1 surface. Surf Sci 181:295–306
17. Fernandez-Torrente I (2008) Local spectroscopy of bi-molecular assemblies: screening, charge transfer, and magnetism at the molecular scale. Ph.D. thesis, FU Berlin
18. Li S, Yu A, Toledo F, Han Z, Wang H, He HY, Wu R, Ho W (2013) Rotational and vibrational excitations of a hydrogen molecule trapped within a nanocavity of tunable dimension. Phys Rev Lett 111:146102
19. Natterer FD, Patthey F, Brune H (2013) Distinction of nuclear spin states with the scanning tunneling microscope. Phys Rev Lett 111:175303

20. Heinrich AJ, Gupta JA, Lutz CP, Eigler DM (2004) Single-atom spin-flip spectroscopy. Science 306:466–469
21. Hamers R (1989) Atomic-resolution surface spectroscopy with the scanning tunneling microscope. Annu Rev Phys Chem 40:531–559
22. Hla S-W (2005) Scanning tunneling microscopy single atom/molecule manipulation and its application to nanoscience and technology. J Vac Sci Technol B 23:1351–1360
23. Image used with permission from Wolfgang Krenner and Peter Weber
24. SPECS GmbH, Voltastrasse 5, 13355 Berlin (DE). http://www.specs.de/cms/upload/PDFs/SPECS__pekte/2010_2_JT-STM_brochure_final_web.pdf
25. Henzler M, Göpel W (1991) Surface science of the solid. B. G. Teubner Stuttgart
26. Zhang L, Miyamachi T, Tomanić T, Dehm R, Wulfhekel W (2011) A compact sub- Kelvin ultrahigh vacuum scanning tunneling microscope with high energy resolution and high stability. Rev Sci Instrum 82:103702
27. Pfeiffer Vacuum Compact Cold Cathode Gauge Typ IKR 270, Berliner Straße 43, 35614 Asslar (DE). https://www.pfeiffer-vacuum.com/de/
28. Oerlikon Leybold DIVAC 2.5T
29. Oerlikon Leybold Vacuum TURBOVAC SL 80
30. Oerlikon Leybold MAG series (turbomolecular pump), Leybold GmbH, Bonner Strasse 498, 50968 Köln (DE). https://www.leybold.com/de/
31. SRS RGA 300, Stanford Research Systems, Inc, 1290-D Reamwood Avenue, Sunnyvale, CA 94089 (USA). http://www.thinksrs.com/products/RGA.htm
32. Cryovac GmbH, Heuserweg 14, 53842 Troisdorf (DE). http://www.cryovac.de
33. Ferrovac GmbH, Thurgauerstrasse 72, 8050 Zürich (CH). http://www.ferrovac.com/
34. Pan S (1993) Piezo-electric motor. Int Pat Publ Number WO 93:19494
35. Cernox temperature sensor, Lake Shore Cryotronics Inc, 575 McCorkle Blvd, Westerville, OH 43082 (USA). http://www.lakeshore.com/Pages/Home.aspx
36. Variable Gain Low Noise Current Amplifier DLPCA-200, FEMTO Messtechnik GmbH, Klosterstraße 64, 10179 Berlin (DE). http://www.femto.de/de/
37. SPECS Zurich GmbH, Technoparkstrasse 1, Zurich (CH). http://www.specs-zurich.com/en/SPM-Control-System.html
38. Ametek lock-in model 7270, Ametek SIGNAL RECOVERY. http://www.ameteksi.com/
39. SPECS Zurich GmbH. http://www.specs-zurich.com/en/SPM-Control-System.html
40. Surface Preparation Laboratory (SPL), Penningweg 69-F, 1507 DE Zaandam (NL). https://www.spl.eu
41. Diller K et al (2013) Free-base and metalated porphyrins on metal surfaces-a systematic X-ray spectroscopy and density functional theory investigation. Ph.D. thesis, Technische Universität München
42. Prof. Dr. Mario Ruben, Institute of Nanotechnology, Karlsruhe Institute of Technology (KIT), 76344 Eggenstein-Leopoldshafen (DE). https://www.int.kit.edu/ruben.php
43. Alfa Aesar GmbH, Zeppelinstrasse 7b, 76185 Karlsruhe (DE). https://www.alfa.com/de/
44. MaTecK GmbH, Im Langenbroich 20, 52428 Juelich (DE). http://www.mateck.de/
45. Kulawik M, Nowicki M, Thielsch G, Cramer L, Rust H-P, Freund H-J, Pearl TP, Weiss PS (2003) A double lamellae dropoff etching procedure for tungsten tips attached to tuning fork atomic force microscopy/scanning tunneling microscopy sensors. Rev Sci Instrum 74:1027–1030
46. Klappenberger F (2014) Two-dimensional functional molecular nanoarchitectures—complementary investigations with scanning tunneling microscopy and X-ray spectroscopy. Prog Surf Sci 89:1–55
47. Lindgren I (2004) Chemical shifts in X-ray and photo-electron spectroscopy: a historical review. J Electron Spectrosc Relat Phenom 137–140:59–71
48. Stöhr J (1992) NEXAFS Spectroscopy, Springer-Verlag
49. Watts JF, Wolstenholme J (2005) Electron spectroscopy: some basic concepts. Wiley
50. Gelius U (1974) Binding energies and chemical shifts in ESCA. Phys Scr 9:133

51. Stepanow S, Strunskus T, Lingenfelder M, Dmitriev A, Spillmann H, Lin N, Barth JV, Wöll C, Kern K (2004) Deprotonation-driven phase transformations in terephthalic acid self-assembly on Cu(100). J Phys Chem B 108:19392–19397
52. Tait SL, Wang Y, Costantini G, Lin N, Baraldi A, Esch F, Petaccia L, Lizzit S, Kern K (2008) Metal-organic coordination interactions in Fe-terephthalic acid networks on Cu(100). J Am Chem Soc 130:2108–2113
53. Hähner G (2006) Near edge X-ray absorption fine structure spectroscopy as a tool to probe electronic and structural properties of thin organic films and liquids. Chem Soc Rev 35:1244–1255
54. Stöhr J (1999) Exploring the microscopic origin of magnetic anisotropies with X-ray magnetic circular dichroism (XMCD) spectroscopy. J Magn Magn Mater 200:470–497
55. Ade H, Hitchcock AP (2008) NEXAFS microscopy and resonant scattering: composition and orientation probed in real and reciprocal space. Polymer 49:643–675
56. ChemDraw Professional 15.0, CambridgeSoft. http://www.cambridgesoft.com/Ensemble_for_Chemistry/ChemDraw/ChemDrawProfessional/
57. HyperChemTM Release 7.0 (MolecularModeling), Hypercube, Inc. http://www.hyper.com/
58. Dewar MJ, Zoebisch EG, Healy EF, Stewart JJ (1985) Development and use of quantum mechanical molecular models. 76. AM1: a new general purpose quantum mechanical molecular model. J Am Chem Soc 107:3902–3909
59. Humphrey W, Dalke A, Schulten K (1996) VMD: visual molecular dynamics. J Mol Gr 14:33–38

Chapter 3
Silver-Bis-Acetylide Wires

3.1 Introduction

The synthesis and characterization of organometallic polymers via on-surface reactions has become a priority in research fields targeting novel molecular-based materials and nanoelectronic devices (see Sect. 1.3.4). Recently, silver-acetylide compounds, i.e., Ag atoms coordinated to sp-hybridized carbon atoms were reported as selective catalysts for the activation of versatile addition reactions taking place under gentle reaction conditions [1–3] (cf. solution-based organometallic intermediates in Fig. 1.2b). Moreover, DFT calculations on terminal alkyne homocoupling on the Ag(111) surface suggest the presence of weak substrate-bound organosilver intermediates prior to the formation of butadiyne bridges (cf. Fig. 4.1b) [4]. A recent study on dehalogenative homocoupling on Au(111) reported on the on-surface formation of organometallic Au-bis-acetylide structures (C(sp)–Au–C(sp) linkage), which convert into covalent polymer structures (C(sp)–C(sp) linkage) [5]. This indicates that organometallic bonding plays a key role in the on-surface reaction pathway toward the creation of well-defined covalent scaffolds.

This chapter focuses on cyano-functionalized metallopolymer wires stabilized through Ag-bis-acetylide bonding on the Ag(110) surface. Specifically, the nanowires are created through the heat-induced on-surface reaction of a customized terminal alkyne linker equipped with polar side groups (CN-DETP). In addition to LT-STM results on the monomer and polymer phase expressing intermolecular interactions through cyano-mediated CN$\cdots$H recognition, C1s XS indicate the formation of organometallic silver–carbon bonding, and reveal the preferential alignment of the Ag-bis-acetylide wires on Ag(110).

R. Hellwig, *Alkyne-Based Nanostructures on Silver Substrates*, Springer Theses,
https://doi.org/10.1007/978-3-030-00997-7_3

3.2 Supramolecular Organization of Alkyne Linker

The chemical structure of CN-DETP synthesized by Ping Du (Ruben group [6]) is presented in the left part of Fig. 3.1a. The quality of the obtained compound was verified through ^{1}H nuclear magnetic resonance (NMR), ^{13}C-NMR, infrared spectroscopy and matrix-assisted laser-desorption-ionisation with time of flight analysis (MALDI-TOF) mass spectroscopy. The linker is comprised of a para-terphenyl backbone with meta-substituted cyano (CN) groups and para-substituted ethynyl moieties. According to previous results, the molecular structure should feature chiral isomerization [7, 8] in combination with alkyne coupling reactions [9]. The surface-adsorbed monomers are subdivided into a prochiral D-trans and L-trans species, and an achiral cis species. These isomer species are represented by atomic ball-and-stick models (modeled via Hyperchem software [10]) in the middle and right part of Fig. 3.1a. During molecular deposition at a substrate temperature $T \leq 250$ K, the cis-trans isomerization barrier is overcome by 180° rotations of the outer phenyl rings, thereby enabling chiral intermolecular adaptation during supramolecular self-assembly. At higher temperatures ($T \geq 300$ K), this reversible process is accompanied by the intermolecular linkage of terminal alkyne monomers through bridging Ag atoms. A gas-phase DFT model for an organometallic nanochain is presented in Fig. 3.1b (Ag atoms within C(sp)–Ag–C(sp) motifs are marked as yellow spheres). The trimer structure was calculated in the framework of the quantum chemistry program ORCA [11] (at the RPBE / def2-SVP level) in combination with the Avogadro software [12].

Since the reactive alkyne moieties are prone to uncontrolled coupling reactions, OMBE experiments under UHV conditions are often limited to a few depositions until the material in the evaporation cell becomes entirely polymerized. For this

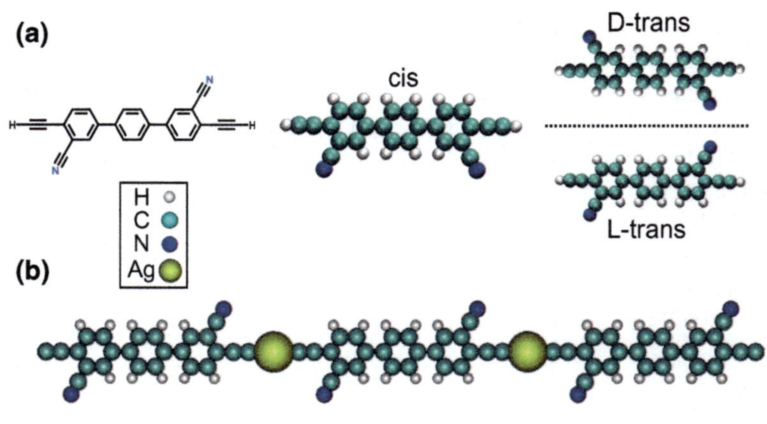

Fig. 3.1 **a** Chemical structure (left) of pristine monomer and ball-and-stick models (middle and right part) of gas-phase configuration. The labeling indicates the isomeric states of CN-DETP. **b** Gas-phase DFT model of organometallic Ag-bis-acetylide trimer. The cyano-phenyl groups can rotate around σ bonds during the assembly of molecular nanochains

reason, it is necessary to verify the chemical integrity of the alkyne monomer prior to initiating on-surface coupling reactions. Fortunately, polymerization processes in the evaporation cell are remarkably slow for CN-DETP compared to DETP (without CN side groups), i.e., multiple powder sublimations can be performed without significant change in molecular flux. From our earlier experience with alkyne derivatives, we assume that steric and electrostatic repulsion between CN side groups efficiently shield the alkyne end groups from coupling reactions within the evaporation cell.

To explore the supramolecular assembly of CN-DETP monomers in the absence of alkyne-based reactions, the molecules were sublimated from an evaporation cell ($T_{\text{sublimation}} = 500$ K) and deposited on the clean Ag(110) substrate kept at 240 K. After cooling the sample down to STM imaging conditions at $T = 4.4$ K, at which monomer diffusion and phenyl rotations are quenched, we observe the formation of 1D ribbons along two perpendicular directions. As shown in Fig. 3.2a, monomers are identified as rod-like protrusions.

From the enhanced and color-coded contrast of the low-bias STM images in Fig. 3.2b, c, we observe depressions for the outer molecules (red arrows) terminating self-assembled 1D ribbons. To isolate an outer molecule, the STM tip was approached until reaching $I_t = 55$ nA on Ag(110) followed by guiding the monomer under OFL conditions along the trajectory of the moving tip (purple line in Fig. 3.2b). From the subsequent STM image in Fig. 3.2c, we conclude that the depressions within the STM contrast coincide with unbound CN moieties. According to the superimposed molecular model, a L-trans monomer is identified. A similar depression-like STM feature was reported by Vitali et al. for a molecule with deprotonated carboxylate and pyridine functionalization on the Cu(111) surface. It was explained by modulations of the tunneling barrier height due to a surface-induced dipole distribution in the vicinity of the electronegative moieties [13]. In our context, chemical sensitivity can be attributed to the STM tip resolving unbound CN moieties, thereby presenting isomer states of surface-adsorbed monomers. Accordingly, the absence of depression-like features next to interacting CN groups implies their engagement in intermolecular interactions.

From the location of the depressions found at the terminations of each ribbon, we conclude that all its conformers express the same isomeric state. Accordingly, all ribbons with positive (negative) inclination with respect to the [001] direction (horizontal line of white cross) are formed by L-trans (D-trans) enantiomers, respectively. Occasionally, the termination molecule of one ribbon binds to another ribbon (Fig. 3.2c). In this case, the latter does not show any depression, which implies the presence of an achiral cis monomer without unbound CN groups.

For a higher molecular coverage (Fig. 3.2d), the ribbon phase gives rise to a maze-like nanostructure. The STM image in Fig. 3.2e highlights that each corner of a rectangular contour (red outline) occurs at the intersection of one L- and one D-trans ribbon, enclosing an angle of 93.6°. From the high-resolution STM topograph in Fig. 3.2f, we conclude that D-trans (L-trans) molecules are slightly tilted (±4°) with respect to the [001] direction (mirror axis depicted as dashed arrow).

Previously, Marschall et al. reported the supramolecular assembly of the related linker [1, 1′; 4′, 1″]-terphenyl-3, 3″-dicarbonitrile, which does not contain terminal

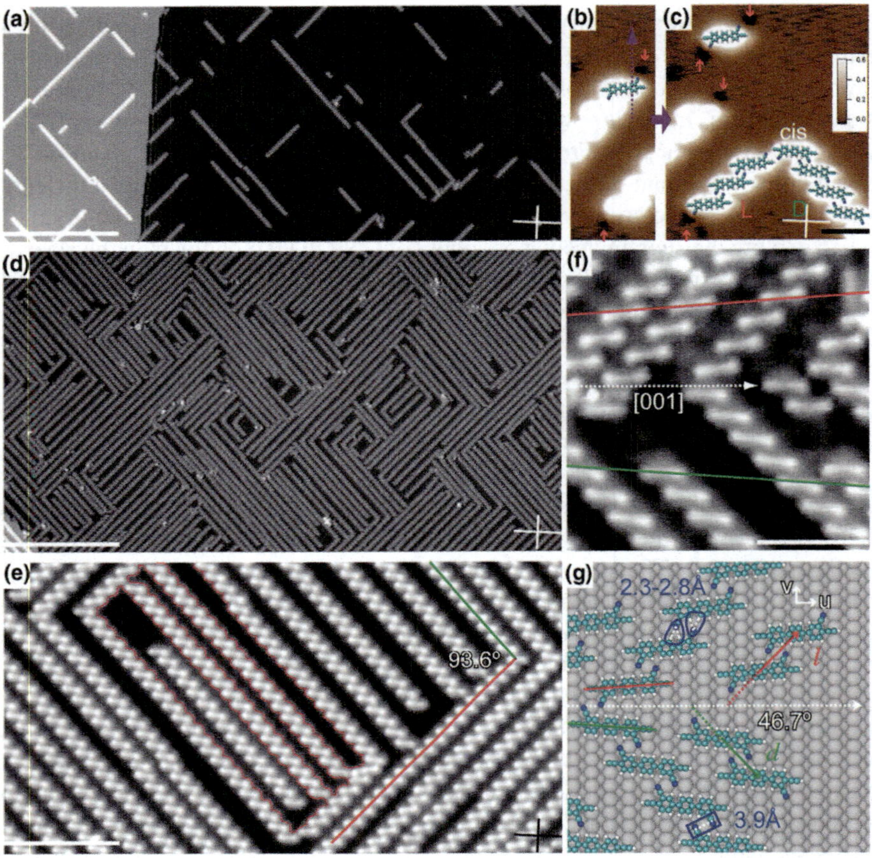

Fig. 3.2 Pristine monomer phase with low (**a**–**c**) and high (**d**–**f**) molecular coverage deposited at 240 K. **a** STM image showing the stacking of monomers along two perpendicular propagation directions. **b**, **c** STM images before and after tip-induced isolation of one monomer, identified as L-trans unit according to CN-related depressions (red markers) and superimposed molecular models. **d**, **e** High-coverage preparation showing maze-like arrangement of D- and L-trans ribbons enclosing rectangular voids. **f** High-resolution STM image reveals molecular alignment (colored lines). **g** Adsorption model for the D-trans and L-trans ribbons. Monomer alignment (colored lines), surface unit cell vectors ($\vec{u}$ and $\vec{v}$) and molecular unit cell vectors ($\vec{d}$ and $\vec{l}$, each enclosing a 46.7° angle with [001] direction (horizontal dashed line)) are indicated. CN$\cdots$H and CN$\cdots$CN distances (numeric values) according to dashed white lines within blue outlines. Tunneling parameters V_t, I_t: **a** 1 V, 0.1 nA; **b**, **c** 0.01 V, 1 nA; **d** 0.1 V, 0.1 nA; **e** −0.1 V, 0.1 nA; **f** 0.01 V, 0.1 nA. Scale bars: **a** 40 nm; **b**, **c** 2 nm; **d** 40 nm; **e**10 nm; **f** 4 nm

alkyne groups [7]. For sub-ML coverages on Ag(111), a similar ribbon phase was observed, but with six molecular orientations (three pairs of D- and L-trans isomers) according to the conformer's mirror symmetry with respect to the three dense-packed crystallographic directions of Ag(111). On the basis of theory calculations, Abbasi-Pérez et al. rationalized the growth of extended D-trans (L-trans) ribbons on Ag(111) by a favorable double hydrogen bonding scenario combined with a particular isomerization mechanism [14, 15].

Consistent with our experiment, the cis → trans isomerization barrier for CN-phenyl rotations is overcome in order to optimize molecular recognition between planar trans enantiomers. Specifically, the regular stacking of the molecules is based on self-complementary CN$\cdots$phenyl interactions according to the so-called PARI [16]. From the STM-based model in Fig. 3.2g, we postulate a commensurate adsorption configuration, which is concluded from the STM image of Fig. 3.3b (explained below). According to the depicted molecular models, the monomers are stacked along the perpendicular "propagation" vectors $\vec{d}$ and $\vec{l}$ of the ribbons, which read

$$\vec{d} = \begin{pmatrix} 2 & -3 \\ 0 & 0 \end{pmatrix} \begin{pmatrix} \vec{u} \\ \vec{v} \end{pmatrix} \text{ and}$$

$$\vec{l} = \begin{pmatrix} 2 & 3 \\ 0 & 0 \end{pmatrix} \begin{pmatrix} \vec{u} \\ \vec{v} \end{pmatrix}$$

with respect to the unit cell vectors $\vec{u}$ and $\vec{v}$ of Ag(110). The vectors are shown in Fig. 3.2g together with the measured CN$\cdots$H and CN$\cdots$CN distances related to electrostatic interactions. The self-complementary stacking via CN$\cdots$phenyl interactions along with the molecular alignment of D- and L-trans conformers (colored lines and 46.7° angle between each colored vector and horizontal dashed line in Fig. 3.2g) rationalize the enclosed angle of 93° between D-trans and L-trans ribbons in Fig. 3.2e.

3.3 Metallopolymer Phase

To explore the potential of surface-confined monomers toward selective alkyne coupling reactions, we thermally annealed the sample with the pristine phase at RT for 10–15 minutes (min). Besides the 1D monomer ribbons with propagation vectors $\vec{d}$ and $\vec{l}$ (cf. Fig. 3.2g), LT-STM measurements reveal motifs not observed before (rectangular outlines in Fig. 3.3a). The STM topograph in Fig. 3.3b displays how the molecules of the ribbons participate in head-to-tail bonding structures embedding sphere-shaped protrusions along the molecular axis. From a careful inspection of the high-resolution STM data in Fig. 3.3c, the bonding distance appears larger than for covalent linkage via butadiyne bridges [9]. Based on the good agreement with the superimposed gas-phase DFT model for a Ag-bis-acetylide trimer (see Fig. 3.1b), we identify STM protrusions between CN-DETP linkers with twofold coordinated Ag

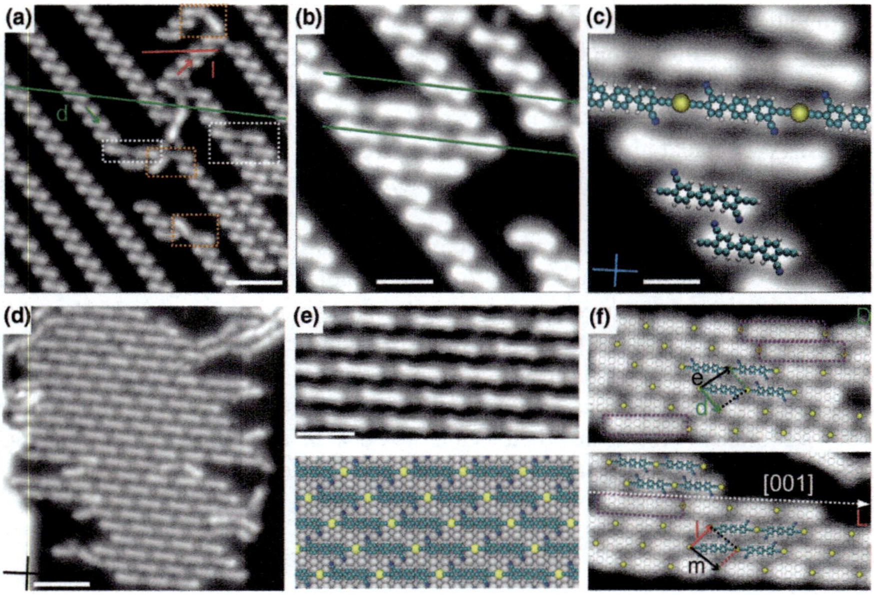

Fig. 3.3 Organometallic structures after sample annealing at 300 K (**a–c**) and 400 K (**d–f**). **a** D- and L-trans ribbons (according to molecular alignment (colored lines) and propagation vectors) coexisting with dimer structures (dashed outlines). **b** Organometallic dimer and trimers connecting two D-trans ribbons. **c** High-resolution STM image with superimposed gas-phase DFT model for organometallic trimer. **d** Dense-packed stacking of 1D silver-bis-acetylide wires within ordered islands. **e** High-resolution STM image of domain (top) and suggested model (bottom). **f** STM images of D- and L-trans domains with superimposed Ag(110) registry and coordinated Ag atoms (yellow). Superimposed molecular models display enantiopure D-trans (L-trans) nanochains and purple outlines depict covalent dimers (byproducts). Compare vectors of molecular unit cell with vectors in (**a**). Symmetry directions are indicated by cross. Tunneling parameters V_t, I_t: **a** −0.1 V, 0.1 nA; **b** 0.1 V, 0.1 nA; **c** 0.01 V, 0.1 nA; **d** 0.05 V, 0.3 nA; **e** 0.005 V, 0.5 nA; **f** 0.01 V, 0.1 nA. Scale bars: **a** 4 nm; **b** 2 nm; **c** 1 nm; **d** 4 nm; **e** 2 nm

atoms (yellow spheres along trimer axis) binding to deprotonated alkynes. For the C–Ag–C coupling motifs, we consider the tendency of Ag atoms to occupy the four-fold hollow sites, in agreement with Ref. [17]. From the observed linkage of adjacent ribbons via Ag-bis-acetylide bridges (Fig. 3.3b), we postulate that unreacted alkynes are located at bridge sites (next to hollow sites), as proposed in the adsorption model in Fig. 3.2g.

A complete conversion from organic ribbons to silver-acetylide chains is accomplished through sample heating at temperatures between 350 and 400 K. STM measurements reveal dense-packed domains (Fig. 3.3d), which display molecular protrusions (bricks) and a periodic lattice of spherical spots. The uniform STM appearance of the interlinked molecules (Fig. 3.3e) indicates that they express the same isomeric state (D-trans). Accordingly, we propose an adsorption model (see bottom part of Fig. 3.3e) with coordinated Ag atoms (yellow) centered at hollow sites of the Ag(110) registry, and deprotonated molecules forming an enantiopure D-trans domain. The coexistent L-trans domain is shown in the bottom part of Fig. 3.3f. From the depicted unit cells of both dense-packed domains separated by a white line along the [001] direction (mirror axis), we recognize the two perpendicular propagation vectors $\vec{d}$ and $\vec{l}$ of the pristine ribbon phase (cf. Figs. 3.3a and 3.2g). The superimposed Ag(110) lattice model shows that all incorporated Ag atoms coincide with fourfold hollow sites. They are assumed to be mobilized through evaporation from kinks and step edges during the annealing process [18]. The linear linking motifs comprised of one Ag atom and two coordinated alkynyl groups are reminiscent of σ-type bonding.

The 2D superlattice of Ag atoms within the domains is rationalized by a CN-mediated interchain recognition between stacked organometallic nanowires. We assume that electrostatic bonding between CN groups and phenyl moieties prevails for the dense-packed network. Since this kind of self-consistent backbone-to-backbone bonding was already observed for the ribbon phase, the formerly introduced vectors $\vec{d}$ and $\vec{l}$ (cf. Fig. 3.2g) are retrieved within the unit cell of the two organometallic domains. This implies that the N atoms do not form metal-organic coordination bonds with Ag atoms, but express supramolecular recognition phenomena.

As apparent from Fig. 3.3c, pristine molecules within the ribbon phase and Ag-coordinated chains show nearly the same alignment. We assume that the regular stacking of linear Ag-acetylide wires is related to the "orthogonality" of C–Ag–C and CN$\cdots$H interactions. Following the previous matrix notation, the D-trans domain expresses the stacking of D-trans wires along the vector $\vec{d}$ according to the matrix notation

$$\begin{pmatrix} \vec{d} \\ \vec{e} \end{pmatrix} = \begin{pmatrix} 2 & -3 \\ 3 & 3 \end{pmatrix} \begin{pmatrix} \vec{u} \\ \vec{v} \end{pmatrix},$$

while the L-trans domain composed of L-trans enantiomers reads

$$\begin{pmatrix} \vec{l} \\ \vec{m} \end{pmatrix} = \begin{pmatrix} 2 & 3 \\ 3 & -3 \end{pmatrix} \begin{pmatrix} \vec{u} \\ \vec{v} \end{pmatrix}.$$

While the alkyne end groups might prefer an exact alignment of C–Ag–C oligomers along the underlying [001] direction [17], CN-mediated interactions may favor a different orientation of the interlinked D- and L-trans isomers (monomers of pristine phase express $\pm 4°$ twist with respect to [001] direction). Accordingly, we assume that the deviation of the chains' alignment from the [001] axis (top panel of Fig. 3.3e) is due to a balance between CN-mediated interchain interactions and site-selective surface anchoring of C–Ag–C bridges. In detail, the distance between two incorporated Ag atoms along the metallopolymer wire amounts to 20.4 Å, which corresponds to five times the lattice constant along the [001] direction of the substrate, indicating the tendency of coordinated Ag atoms to coincide with hollow sites. From the gas-phase DFT calculation on the organometallic trimer in Fig. 3.1b, we encounter a larger distance (21.12 Å) between two coordinated Ag atoms. Thus, for the surface-adsorbed organometallic chains, we suspect that this length difference is compensated by the downward bending of the alkynes toward the surface.

Besides organometallic dimers and trimers closely aligned along the molecular axis of the monomers (white outlines in Fig. 3.3a), a small quantity of dimers expresses either a shorter dimer length (purple outlines in Fig. 3.3f) or a kink motif (orange outlines in Fig. 3.3a). As will be discussed later (cf. Fig. 3.6), these byproducts are assigned to covalent C–C bond formation.

3.4 X-ray Spectroscopy Analysis

After unraveling the adsorption behavior and structural properties of CN-DETP on Ag(110), the chemical aspects of the pristine and organometallic phase are inspected via synchrotron-based XS experiments. On the basis of the STM-based preparation protocols mentioned above, the clean Ag(110) substrate is exposed to a sub-ML amount of molecules at a sample temperature of 200 K and characterized both before and after sample annealing at 400 K. While STM experiments only elucidate microsopic regions of the molecular adlayer, XS-based results present ensemble properties averaged over the macroscopic spot size of the X-ray beam. To reduce monomer mobility and radiation damage, all spectroscopy experiments were conducted at a sample temperature of 190–200 K.

C1s XPS spectra of the pristine sample (ribbon phase) (cf. Fig. 3.2e) and annealed sample (cf. Fig. 3.3d) are presented in the top and bottom part of Fig. 3.4, together with depicted monomer and dimer models referring to a pristine (annealed) molecular adlayer. The combination of the intrinsic core level lifetime, substrate-induced peak broadening and limited experimental resolution does not allow to disentangle each chemical species contributing to the XPS profile (red data points) in Fig. 3.4. The latter is approximated by the sum of individual Voigt fits (black curves) accounting for the different chemical environments of C(sp) within C≡N, C≡C and C(sp^2) units within the monomer backbone. The high BE shoulder at 286.5 eV (blue fitting curve) for the pristine phase (top panel) indicates spectral contributions from CN end groups. The maximum of the XPS profile shifts down from 285.2 to 284.9 eV after sample annealing at 400 K (bottom panel). For the latter, a low BE shoulder emerges at 283.9 eV (orange fitting curve). Within the scope of a previous study on the alkyne derivative TEB, DFT-modeled XPS line shapes predicted the appearance of a low BE shoulder for the organosilver TEB–Ag–TEB dimer, while the covalent TEB–TEB dimer does not exhibit this feature (see Fig. 6b of Ref. [19]). Based on the related DFT-calculation, the experimentally observed low BE shoulder is considered as a spectroscopic evidence for organometallic bonding via Ag-bis-acetylide bridges. The signature is also consistent with literature data, e.g., Kung et al. reported a BE of 283.7 eV for unsaturated methylacetylide compounds [20].

Complementary to XPS, synchrotron-based NEXAFS spectroscopy is utilized to elucidate geometrical, conformational and organometallic bonding aspects within CN-functionalized Ag-bis-acetylide chains. Figure 3.5a depicts the polar NEXAFS configuration, where the sample is rotated toward one of three orientations with respect to the incident X-ray beam (angle φ), i.e., grazing (25°), magical (53°) and

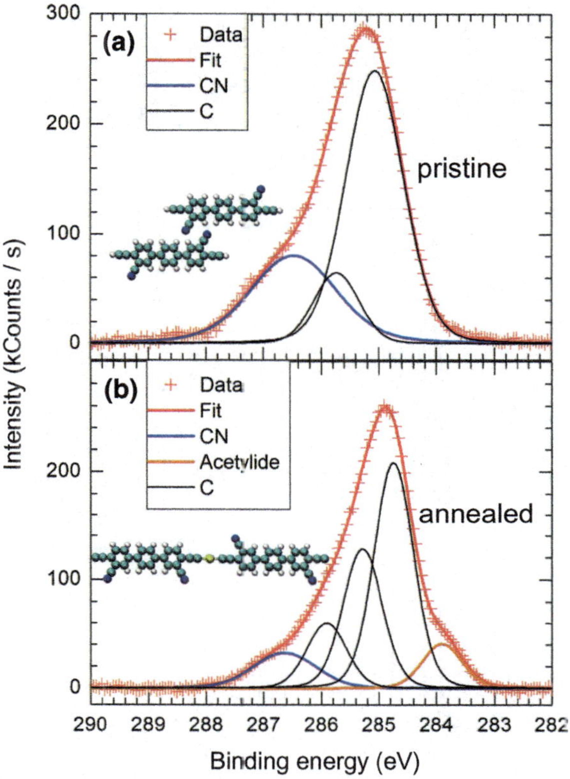

Fig. 3.4 C1s XPS characterization of pristine (**a**) and annealed sample (**b**) (pursuant to depicted molecular models). The comparison reveals a downshift in BE (285.2–284.9 eV) and the emergence of a low-energy shoulder (283.9 eV) upon sample annealing at 400 K

normal (90°) incidence. C1s NEXAFS resolves the π^* orbitals of C(sp) atoms within terminal alkynes (orange orbitals) and CN end groups (purple orbitals), as indicated by the illustrative model showing pristine monomers and reacted Ag-acetylide chains aligned on Ag(110) (Fig. 3.5a, d). They are subdivided into out-of-plane π^* orbitals oriented perpendicular to flat-lying phenyl rings (perspective view on left side) and in-plane π^* orbitals (top view on right side).

The C1s NEXAFS spectra probing the orientation of the out-of-plane orbitals are shown for the pristine (organometallic) phase in Fig. 3.5b, c. The extended profiles display four sharp π^* resonances resolved at the photon energies {284.9, 285.9, 286.8 and 288.6} eV, as labeled by the capitalized letters A, B, C and D. The high photon energy window above 290 eV is indicative for the σ^* region with the characteristic overlapping of broadened σ^* peaks (feature E at 293.3 eV). A qualitative comparison of the three profiles reveals a strong dichroism for the four π^* peaks, while the σ^* region exhibits an opposite angular behavior. Based on comparison with the NEXAFS study by Solomon et al. providing peak positions for a sub-ML amount of benzene on Ag(110) ($\{\pi^*_{1,2}, \pi^*_3, \sigma^*_1$ and $\sigma^*_2\}$ = {284.9, 288.6, 293.4 and 301} eV), the features A and D can be assigned to the out-of-plane π^* orbitals of the C(sp^2) atoms of the

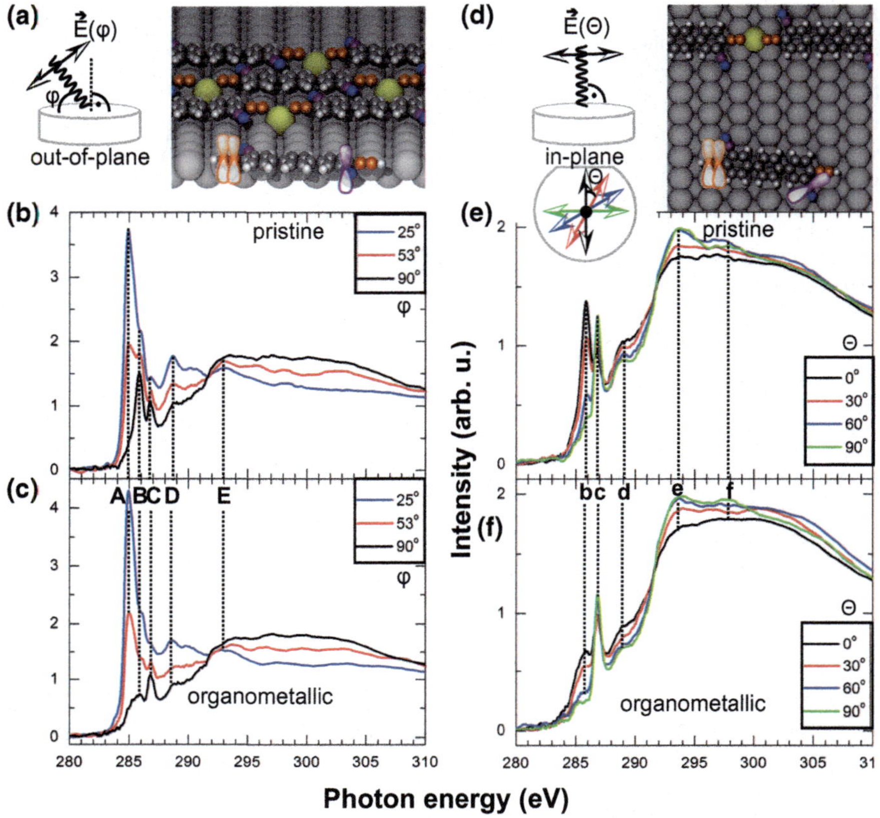

Fig. 3.5 C1s NEXAFS analysis of polar (**a–c**) and azimuthal (**d–f**) angular dependence for the pristine (**b**, **e**) and organometallic phase (**c**, **f**). **a**, **d** Schematic sketch depicting polar (azimuthal) NEXAFS geometry according to variations of out-of-plane (in-plane) angle φ (θ) with respect to electrical field vector $\vec{E}$. Perspective and top view of an illustrative model emphasizing out-of-plane and in-plane π^* orbitals of C(sp) atoms within C≡N and C≡C groups (only shown for monomer). Color-coding: C(sp) within alkynes (CN groups): orange (purple); C(sp^2): dark grey; N: blue; H: white. **b**, **c** Polar NEXAFS dichroism for the pristine phase (above) and organometallic phase (below). **e**, **f** Azimuthal NEXAFS dichroism for the pristine phase (above) and organometallic phase (below). Dashed lines with capitalized (lowercase) letters refer to peak positions of polar (azimuthal) NEXAFS spectra

molecular backbone. For normal incidence, the latter are almost quenched, indicating a nearly planar molecular backbone (cf. Fig. 3.5a).

The features B and C become more prominent with increasing angle φ, thus indicating a significant contribution from in-plane π^* orbitals, which stem from C(sp) atoms. Since the peak position of B is consistent with previous results for a related alkyne derivative [21], and peak position C agrees with prior studies for para-substituted dicarbonitrile-oligophenyls [22], we assign them to excitations with initial states of the C(sp) atoms of the alkyne and CN species, respectively.

To inspect the angular dependence for each peak signature, we divided the peak intensity for grazing incidence ($\varphi = 25°$) by the intensity for perpendicular incidence ($\varphi = 90°$). Comparing this peak ratio for the pristine and annealed sample, the ratios of C, D and E remain nearly constant upon organometallation ($1.34 \rightarrow 1.53$, $1.75 \rightarrow 1.94$ and $0.90 \rightarrow 0.88$). The moderate decrease of the A-ratio ($9.70 \rightarrow 8.75$) indicates slightly less planar molecules for the organometallic chains within dense-packed domains compared to monomers within 1D ribbons. Interestingly, there is a drastic change (factor two) for the B-ratio ($1.44 \rightarrow 3.07$), compared to the nearly unaltered angular dependence of the CN-related resonance C. The strong broadening and diminished peak height of signature B upon organometallation suggests that the alkyne's π^* orbitals hybridized with orbitals of coordinated Ag atoms, while the π^* orbitals of CN-DETP remain nearly unaffected.

In order to probe the preferential alignment of the organometallic chains within the sample plane and to inspect the in-plane π orbitals of CN and alkyne groups after organometallation reactions, further C1s NEXAFS measurments are carried out under normal incidence (azimuthal angular dependence). According to the schematic sketch in Fig. 3.5d, X-rays arrive under perpendicular incidence on the sample (black profiles in Fig. 3.5b, c). Since the $[1\bar{1}0]$ direction along the furrows of the Ag(110) is known (macroscopic edge on crystal), this surface direction was initially aligned with the electrical field vector $\vec{E}$ (angle $\theta = 0°$). By rotating the sample in 30° steps, we probed the directionality of the in-plane orbitals for the pristine and annealed phase. The relative angle θ between $\vec{E}$ and the $[1\bar{1}0]$ direction of the atomic lattice orientation is depicted by color-coded arrows (Fig. 3.5d) that refer to the pertaining C1s NEXAFS profiles of Fig. 3.5e, f. Consistent with the former peak assignment (B-E), the letters b, c, d, e and f represent the pertaining signatures at the photon energies {285.9, 286.9, 289.0, 293.9 and 297.5} eV. The dichroism is quantified by the peak ratio between the $\theta = 0°$ spectrum (black profile) and $\theta = 90°$ curve (green profile). According to the peaks B and D, the angular behavior of b and d is assigned to the in-plane π^* orbitals of the alkyne species (orange coloring).

Since the most intense NEXAFS peak A is absent for $\varphi = 90°$ (Fig. 3.5b), one recognizes that feature b in Fig. 3.5e, f shows the highest signal for $\theta = 0°$, which is explained by a parallel alignment between the electrical field vector and alkyne's in-plane π^* orbital. It implies for both the pristine and reacted phase, that monomers and organometallic wires preferentially align along the [001] direction, which is perpendicular to these orbitals. This conclusion is consistent with our STM results limited to microscopic sample regions. According to its linear bonding geometry, the organometallic chains represent 1D arrays of σ-like metal alkynyl complexes [23] (see Sect. 1.3.4).

From the nearly constant peak ratio of the CN-related signature c ($0.76 \rightarrow 0.85$), we conclude that the conformation of the CN-phenyl moieties and molecular alignment has not significantly changed upon the conversion from 1D ribbons to dense-packed domains. For both pristine and reacted phase, the STM results reveal the presence of self-complementary CN$\cdots$phenyl interactions between isochiral trans isomers (cf. Figs. 3.2c and 3.3e). While LT-STM clearly resolved the enantiomeric character of monomers and oligomers (D- and L-trans), NEXAFS experiments space-average over both chiral species. Since the CN moieties of both D-trans and L-trans isomers confine an angle of 120°, there is no field vector alignment θ for which the contributions from CN groups get quenched.

Compared to the nearly unchanged angular behavior of the CN-related peak c upon sample annealing, the NEXAFS spectra in Fig. 3.5f reveal a drastically attenuated and broadened peak b attribute to the alkynes' in-plane π^* orbitals next to twofold Ag-coordination sites. As for the polar angular dependence in Fig. 3.5b, c, the azimuthal peak ratio of b strongly decreases (3.41 $\rightarrow$ 2.71). Therefore, we explain the drastic change of signature b by a hybridization between in-plane π^* orbitals and orbitals of Ag atoms involved in organometallic C(sp)–Ag–C(sp) bonding. The emergence of a minute amount of covalent byproducts with deviating orientations upon sample annealing is neglected (see explanation below; cf. orange and purple outlines in Fig. 3.3a, f), in agreement with the observation that the angular behavior and peak ratio of the CN-related NEXAFS peaks C and c hardly changed after the on-surface reaction.

3.5 Covalent Byproducts

The relationship between organometallic and covalent terminal alkyne linkage is explored on Ag(110), in order to clarify whether the linear C≡C–Ag–C≡C complex represents an intermediate for covalent C≡C–C≡C bond formation (homocoupling). There exists uncertainty whether the heat-induced split-up of Ag-acetylide chains allows the reversible uptake of another Ag atom (exchange process), bis-acetylide $\rightarrow$ mono-acetylide conversion, or exclusively leads to radical recombination via covalent bonding. Interestingly, after successive annealing steps within the temperature regime 350 K $\leq T \leq$ 600 K, we clearly observe mono-acetylide complexes (C(sp)–Ag motif in Fig. 3.6d), as indicated by the white arrows in Fig. 3.6e–g. This bonding scenario could either be established prior to C(sp)–Ag–C(sp) formation or/and after C(sp)–Ag–C(sp) scission processes. In the latter case, the scission of two C(sp)–Ag–C(sp) motifs may lead to two Ag–C(sp) motifs and two C(sp) radicals forming a covalent dimer when they encounter each other.

The reaction products not attributed to organometallic bonding are presented with a color-coded STM contrast in Fig. 3.6a–c in order to recognize the presence/absence of depressions reflecting unbound CN groups. After room-temperature annealing, we observe a minute concentration of an on-surface formed dimer species with a kink motif, as shown within the orange outlines in Fig. 3.6a, e, f. Based on the good agreement with the superimposed molecular model in Fig. 3.6a, we suggest an enyne motif (see chemical structure in the top part of Fig. 3.6a). Consistently, a further regioisomer emanating from vinyl-ethynyl linkage is encountered, as marked by purple outlines in Fig. 3.6b, e. From the superimposed Ag(110) registry in Fig. 3.3f, it becomes evident that the Ag atoms attached to the left and right end of the covalent dimer (purple outline) are separated by nine lattice constants (36.8 Å) along the [001] direction, which is one lattice constant less compared to an organometallic

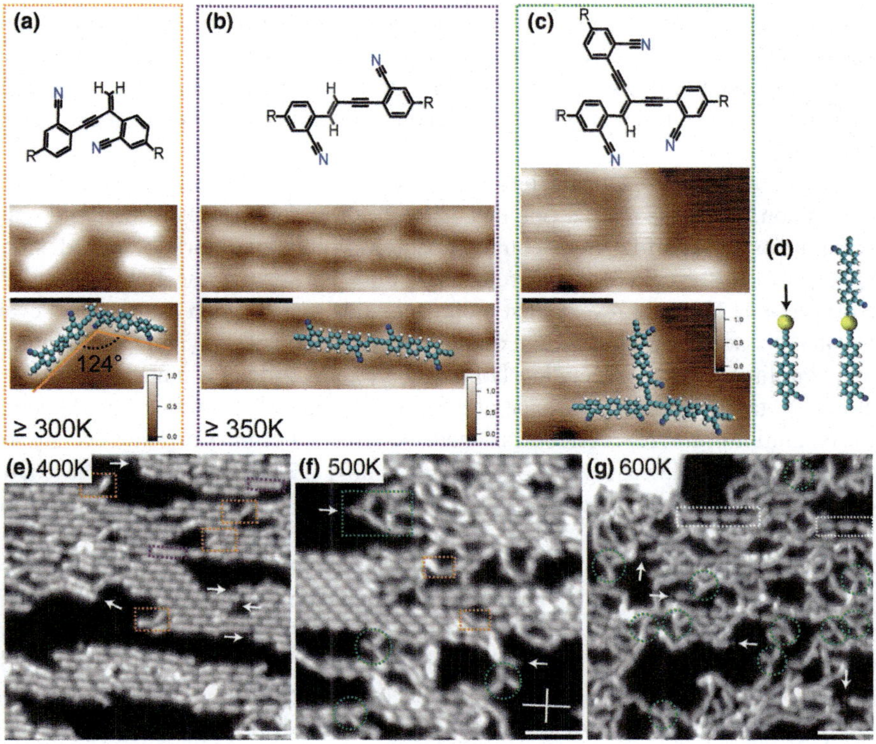

Fig. 3.6 Covalent byproducts. **a–c** Proposed chemical structures (top part) according to STM images of covalent coupling motifs (middle part), along with suggested gas-phase ball-and-stick models (bottom part) superimposed to STM images. Color-coded STM contrast helps in identifying non-interacting CN groups. Within chemical structure, R represents residual structure of CN-DETP monomer. **a, b** Covalent bond formation via vinyl-ethynyl catenation. **c** Covalent trimer. **d** Schematic representation of mono- and bis-acetylide complex. High-temperature sample annealing causes increase of covalent byproducts. **e** 400 K: Organometallic domains with dimer products (orange and purple outline). **f** 500 K: Dissolvement of domains and formation of trifurcated trimer products (green outline). **g** 600 K: Irregular polymer branching with trifurcation motifs and sporadic presence of Ag-bis-acetylide wires (white outline). Tunneling parameters V_t, I_t: **a** −0.002 V, 0.1 nA; **b** 0.005 V, 0.3 nA; **c** 0.05 V, 0.1 nA; **e** 0.01 V, 0.1 nA; **f** 0.8 V, 0.8 nA; **g** 0.01 V, 0.1 nA. Scale bars: **a–c** 2 nm; **e–g** 6 nm

dimer with terminal Ag atoms. The high-resolution STM image of the covalent dimer motif (formed at $T \geq 350$ K) in Fig. 3.6b perfectly agrees with the superimposed gas-phase model, i.e., a small kink associated with vinyl-ethynyl catenation (see chemical structure) is clearly expressed. Therefore, we conclude the absence of terminal alkyne homocoupling.

Threefold coupling motifs are rare for temperatures $T \leq 400$ K, but become more frequent for sample annealing at $T \geq 500$ K (green outline in Fig. 3.6f, g). For the species shown in Fig. 3.6c, a threefold coupling motif expressing vinyl-ethynyl-ethynyl linkage is suggested (see depicted chemical structure). While only a minute amount of enyne byproducts (orange and purple outline in Fig. 3.6e) forms after sample annealing at 400 K, sample heating at 500 K causes the gradual dissolution of dense-packed aggregates (Fig. 3.6f). The structure conversion at domain borders is characterized by irregularly branched chains with threefold coupling nodes (green outline).

Subsequent to sample annealing at 600 K (Fig. 3.6g), we observe the complete conversion into a disordered polymer network. The dominance of threefold coupling nodes and fourfold cross-coupling patterns within irregularly branched structures [24] may be caused by interchain reactions between Ag-acetylide motifs. A closer look in Fig. 3.6g reveals that mono-acetylide motifs (white arrows) and even isolated bis-acetylide wires (white outline) are preserved, thus evincing a high thermal stability of the Ag-acetylide bonding between CN-DETP molecules and Ag atoms. Notably, the irregular shape of the covalent on-surface products starkly contrasts solution-based reaction routes, where metal-acetylide intermediates occur prior to the formation of linear diyne motifs (cf. Fig. 1.2b) [1, 25]. Consequently, the results exemplify that an ordered organometallic structure does not necessarily represent the intermediate toward a topologically equivalent covalent framework. Within a related study reporting on the on-surface coupling of acetylene molecules toward Cu-bis-acetylide chains on Cu(110), high-temperature annealing also led to the degradation of the linear structure instead of conversion into linear covalent strands [26]. According to the DFT-calculated homocoupling mechanism on Ag(111) of Björk et al. predicting that the coupling step precedes the deprotonation step [4], the absence of homocoupling in our case could be related to alkyne deprotonation prior to intermolecular coupling on Ag(110).

3.6 Conclusion

In summary, the bottom-up creation and multi-technique characterization of CN-functionalized Ag-acetylide chains is presented. To this end, we employed the de novo synthesized precursor CN-DETP, which is comprised of terminal alkyne end groups and CN side groups. Firstly, LT-STM is used to study the supramolecular self-assembly of the pristine monomer on Ag(110). The growth of 1D ribbons along two directions was ascribed to the anisotropic registry of Ag(110) along with weak electrostatic interactions between CN groups and phenyl moieties of the backbone. The isomeric state of the monomers is clearly resolved from CN-related depressions within the STM contrast. Following dehydrogenative coupling on Ag(110), already annealing at RT gives rise to chain-like organometallic motifs expressing C(sp)–Ag–C(sp) linking. For sample annealing at 400 K, we observe the complete conversion from the monomer phase to enantiopure 2D domains of regularly stacked Ag-bis-acetylide molecular wires. As for the monomer phase, CN-mediated interactions cause the periodic stacking of trans isomers. Besides the predominance of silver-bis-acetylide linkage, a minute amount of silver-mono-acetylide and covalent motifs is identified for heat-induced reactions at $T \leq 400$ K. The covalent products increase for annealing at higher sample temperatures, while homocoupling remains absent. Annealing at 600 K does not lead to covalent nanochains, but disordered polymer structures that coexist with a few isolated Ag-bis-acetylide wires. The latter indicates a high thermal stability for organometallic silver alkynyl bonding within 1D structures. The monomer phase and organometallic phase (for sample annealing at $T \leq 400$ K) are further characterized through XS measurements. C1s XPS results deliver a clear spectroscopic fingerprint for organometallic carbon–silver bonding on Ag(110). By means of NEXAFS measurements with varying polar and azimuthal angles, the preferential alignment of the linear Ag-bis-acetylide chains is determined. In addition to structural details, NEXAFS peaks from sp-hybridized π^* orbitals within functional C≡C groups reveal significant changes upon sample annealing at 400 K, which we attribute to hybrizidation between Ag orbitals and in-plane π^* orbitals of the alkynes. The results provide an approach toward the on-surface synthesis of thermally stable and CN-functionalized silver alkynyl chains. Due to selective isomerization processes following supramolecular side recognition principles, the organometallic chains arrange in a regular fashion representing a periodic superlattice of twofold-coordinated Ag atoms.

References

1. Halbes-Letinois U, Weibel J-M, Pale P (2007) The organic chemistry of silver acetylides. Chem Soc Rev 36:759–769
2. Yamamoto Y (2008) Silver-catalyzed C_{sp}-H and C_{sp}-Si bond transformations and related processes. Chem Rev 108:3199–3222
3. Fang G, Bi X (2015) Silver-catalyzed reactions of alkynes: recent advances. Chem Soc Rev 44:8124–8173
4. Björk J, Zhang Y-Q, Klappenberger F, Barth JV, Stafstrm S (2014) Unraveling the mechanism of the covalent coupling between terminal alkynes on a noble metal. J Phys Chem C 118:3181–3187
5. Sun Q, Cai L, Ma H, Yuan C, Xu W (2016) Dehalogenative homocoupling of terminal alkynyl bromides on Au(111): incorporation of acetylenic scaffolding into surface nanostructures. ACS Nano 10:7023–7030
6. Prof. Dr. Mario Ruben, Institute of Nanotechnology, Karlsruhe Institute of Technology (KIT), 76344 Eggenstein-Leopoldshafen (DE). https://www.int.kit.edu/ruben.php
7. Marschall M, Reichert J, Weber-Bargioni A, Seufert K, Auwrter W, Klyatskaya S, Zoppellaro G, Ruben M, Barth JV (2010) Random two-dimensional string networks based on divergent coordination assembly. Nat Chem 2:131–137
8. Marschall M, Reichert J, Seufert K, Auwrter W, Klappenberger F, Weber-Bargioni A, Klyatskaya S, Zoppellaro G, Nefedov A, Strunskus T, Wöll C, Ruben M, Barth JV (2010) Supramolecular organization and chiral resolution of p-terphenyl-mdicarbonitrile on the Ag(111) surface. Chem Phys Chem 11:1446–1451
9. Cirera B, Zhang Y-Q, Björk J, Klyatskaya S, Chen Z, Ruben M, Barth JV, Klappenberger F (2014) Synthesis of extended graphdiyne wires by vicinal surface templating. Nano Lett 14:1891–1897
10. HyperChem™ Release 7.0 (MolecularModeling), Hypercube, Inc. http://www.hyper.com/
11. Neese F (2012) The ORCA program system Wiley Interdiscip. Rev Comput Mol Sci 2:73–78

12. Hanwell MD, Curtis DE, Lonie DC, Vandermeersch T, Zurek E, Hutchison GR (2012) Avogadro: an advanced semantic chemical editor, visualization, and analysis platform. J Cheminf 4:17
13. Vitali L, Levita G, Ohmann R, Comisso A, Vita AD, Kern K (2010) Portrait of the potential barrier at metal-organic nanocontacts. Nat Mater 9:320–323
14. Abbasi-Pérez D, Recio JM, Kantorovich L (2014) Building motifs during self-assembly of para-terphenyl-meta-dicarbonitrile on a metal surface: a gas-phase study. J Phys Chem C 118:10358–10365
15. Abbasi-Pérez D, Manuel Recio J, Kantorovich L (2015) The role of isomerization in the kinetics of self-assembly: p-terphenyl-m-dicarbonitrile on the Ag(111) surface. Phys Chem Chem Phys 17:11182–11192
16. Arras E, Seitsonen AP, Klappenberger F, Barth JV (2012) Nature of the attractive interaction between proton acceptors and organic ring systems. Phys Chem Chem Phys 14:15995–16001
17. Liu J, Chen Q, Xiao L, Shang J, Zhou X, Zhang Y, Wang Y, Shao X, Li J, Chen W, Xu GQ, Tang H, Zhao D, Wu K (2015) Lattice-directed formation of covalent and organometallic molecular wires by terminal alkynes on Ag surfaces. ACS Nano 9:6305–6314
18. Zambelli T, Barth JV, Wintterlin J (1998) Formation mechanism of the O-induced added-row reconstruction on Ag(110): a low-temperature stm study. Phys Rev B 58:12663–12666
19. Zhang, Y-Q Kepčija N, Kleinschrodt M, Diller K, Fischer S, Papageorgiou AC, Allegretti F, Björk J, Klyatskaya S, Klappenberger F, Ruben M, Barth JV (2012) Homocoupling of terminal alkynes on a noble metal surface. Nat Commun 3:1286
20. Kung H, Wu S-M, Wu Y-J, Yang Y-W, Chiang C-M (2008) Tracking the chemistry of unsaturated C_3H_3 groups adsorbed on a silver surface: propargyl-allenyl-acetylide triple bond migration, self-hydrogenation, and carbon–carbon bond formation. J Am Chem Soc 130:10263–10273
21. Zhang Y-Q, Björk J, Weber P, Hellwig R, Diller K, Papageorgiou AC, Oh SC, Fischer S, Allegretti F, Klyatskaya S, Ruben M, Barth JV, Klappenberger F (2015) Unusual deprotonated alkynyl hydrogen bonding in metal-supported hydrocarbon assembly. J Phys Chem C 119:9669–9679
22. Klappenberger F, Kuhne D, Marschall M, Neppl S, Krenner W, Nefedov A, Strunskus T, Fink K, Wöll C, Klyatskaya S, Fuhr O, Ruben M, Barth JV (2011) Uniform p-system alignment in thin films of template-grown dicarbonitrile-oligophenyls. Adv Funct Mater 21:1631–1642
23. Long NJ, Williams CK (2003) Metal alkynyl σ complexes: synthesis and materials. Angew Chem Int Ed 42:2586–2617
24. Eichhorn J, Heckl WM, Lackinger M (2013) On-surface polymerization of 1,4-diethynylbenzene on Cu(111). Chem Commun 49:2900–2902
25. Siemsen P, Livingston RC, Diederich F (2000) Acetylenic coupling: a powerful tool in molecular construction. Angew Chem Int Ed 39:2632–2657
26. Sun Q, Cai L, Wang S, Widmer R, Ju H, Zhu J, Li L, He Y, Ruffieux P, Fasel R (2016) Bottom-up synthesis of metalated carbyne. J Am Chem Soc 138:1106–1109

Chapter 4
Fabrication of Graphdiyne Nanowires

The main results from this chapter were published in the journal "Small" (see chapter "List of publications").

4.1 Introduction

Beyond metal alkynyl nanostructures, mixed sp–sp^2 covalent organic framework (COF) such as GDY are believed to express extraordinary structural, mechanical and optoelectronic properties (see Sect. 1.3.2). On-surface homocoupling of terminal alkyne monomers is a promising approach to synthesize GDY-like structures (see Sect. 1.3.5). However, the first attempts targeting the synthesis of extended GDY nanowires via the coupling of the alkyne linker DETP on Ag(111) resulted in diverse byproducts (see Fig. 4.5a). The linear growth mode toward GDY, i.e., an alternating sequence of $n \geq 1$ phenyl rings and one diyne unit, was improved by utilizing a vicinal surface with closely spaced step edges (Fig. 1.5b, c) [1].

In the last chapter, we have shown that the molecule CN-DETP (see Fig. 4.1a) deposited on the Ag(110) surface leads to the formation of organometallic nanowires expressing high thermal stability. Here, we study on-surface reactions with CN-DETP on the Ag(111) surface facilitating covalent C(sp)–C(sp) linkage instead of C(sp)–Ag–C(sp) bonding. According to DFT-based calculations, the terminal alkyne homocoupling reaction pathway passes through several intermediate steps, i.e., vinyl–vinyl and vinyl–ethynyl linkage motifs with carbon–silver bonds that are removed upon deprotonation (Fig. 4.1b) [2]. The only byproduct that we encounter for the homocoupling of CN-DETP is an enyne motif (vinyl–ethynyl catenation). We succeeded in isolating GDY-like polymer wires and characterizing their mechanical and conformational characteristics.

R. Hellwig, *Alkyne-Based Nanostructures on Silver Substrates*, Springer Theses,
https://doi.org/10.1007/978-3-030-00997-7_4

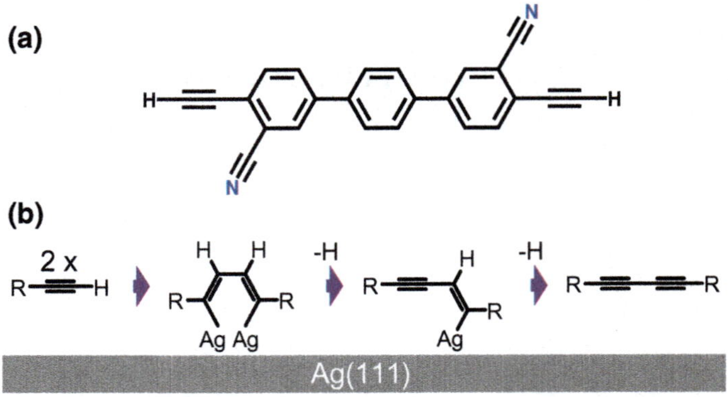

Fig. 4.1 **a** Chemical structure of CN-DETP. **b** On-surface reaction mechanism of homocoupling (R represents residue group)

4.2 Alkyne-Based Tecton

After depositing a sub-ML of CN-DETP ($T_{sublimation}$ = 460 K) on the clean Ag(111) surface kept at $T \leq 300$ K, transferring the sample to the STM and cooling down to 4.5 K, the monomer phase of CN-DETP on Ag(111) is characterized. All LT-STM topographs in this chapter are presented with a brownish STM contrast in order to enhance the visibility of CN-related depressions close to unbound cyano (CN) moieties, which we observe for CN-DETP on Ag(110) (see explanation in Chap. 3, Fig. 3.2b, c), Ag(111) and Ag(100). As depicted by white arrows (Fig. 4.2a), depressions are discerned at edge sites, where one CN group points outwards for each molecule. Therefore, we suggest the presence of enantiopure domains comprised of trans enantiomers, as indiciated by the superimposed gas-phase models (created via Hyperchem [3]) in Fig. 4.2b. From the latter along with the depicted molecular unit cell (green outline), one sees that the linkers express head-to-tail bonding. In contrast to the 1D ribbon phase on Ag(110), stabilized through CN$\cdots$phenyl interactions, the dense-packed phase on Ag(111) is rationalized by additional attractions between the CN and alkyne species. Within the alkyne-based supramolecular tecton shown in the upper part of Fig. 4.2c (orange outline, each alkyne (orange) is engaged in self-complementary C≡C–H$\cdots$N≡C bonding with an opposite CN group (cyan). The CN$\cdots$phenyl bonding with the molecular backbone (PARI [4]) is marked by the yellow outline in the lower part of Fig. 4.2c. The X-ray crystal structure analysis of a CN-DETP molecular crystal (obtained by slowly evaporating its solution in EtOAc) confirms the self-complementary character of both motifs and delivers structural parameters. In the visualization model in Fig. 4.2d, the vectors $\vec{a}$, $\vec{b}$ and $\vec{c}$ span a triclinic unit cell with molecules stacked along the vector $\vec{a}$. The distance D within the alkyne tecton (2.48 Å) indicates weak electrostatic bonding, while DFT-calculations on a 2-ethynyl-benzonitrile molecule cluster predict a N$\cdots$H distance of 2.2 Å and BE of 0.3 eV. The N$\cdots$H distances d = 2.73 and d' = 2.75 Å for the 3D

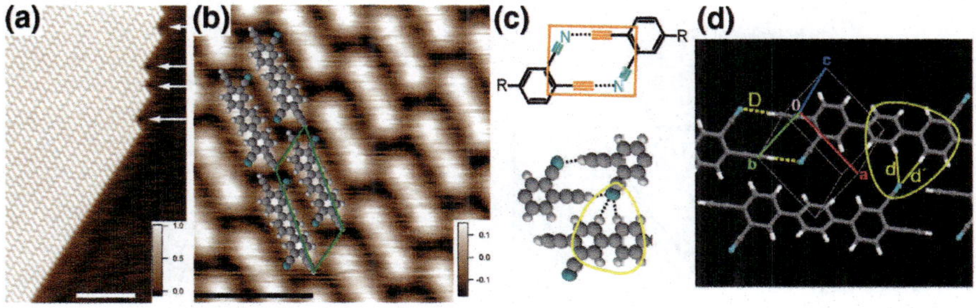

Fig. 4.2 Monomer phase of CN-DETP on Ag(111). **a**, **b** STM images showing organic domain. Some edge molecules in (**a**) express one CN-related depression (arrows). Gas-phase models and molecular unit cell are superimposed in (**b**). Hydrogen: white; carbon: grey; nitrogen: cyan. **c** Head-to-tail bonding between alkyne and CN moieties (orange outline) and CN··· phenyl head-to-backbone interactions (yellow outline). **d** Visualization from X-ray crystal structure data for CN-DETP crystal. Color-coded unit cell vectors ($\vec{a}$, $\vec{b}$ and $\vec{c}$) as well as N··· H distances D (dashed line), d and d' (solid lines within contour) are indicated. Tunneling parameters V_t, I_t: **a** −500 mV, 0.1 nA; **b** −500 mV, 0.05 nA. Scale bars: **a** 8 nm; **b** 2 nm

crystal are typical values for weak CN-mediated electrostatic interactions. The self-complementary character of the CN-alkyne tecton opens up new synthesis protocols toward 3D molecular crystal engineering based on two different sp-functionalized groups.

4.3 Molecular Desorption Via STM Tip

Inspecting the electronic properties of the monomer phase, we performed *dI/dV* spectroscopy. In detail, Fig. 4.3a shows STM images before and after acquiring a STS spectrum at the depicted spot (red-black marker) within a monomer island. The forward spectrum (fwd) from 1 to 2.5 V indicates the lowest unoccupied molecular orbital (LUMO) state at 1.28 V and onset of the LUMO+1 peak, which is followed by a drop in the *dI/dV* signal above 2.23 V (green-colored region of spectrum). The backward (bwd) spectrum (red profile) shows only one STS peak within this voltage interval, i.e., the LUMO+1 peak appears with a reduced intensity and shift to 2.2 Vl, while the LUMO peak vanished. This signal drop in the fwd spectrum together with a different spectral fingerprint in the bwd spectrum is conspicuous for an abrupt change in the STM junction. Accordingly, the subsequent STM image in the lower part of Fig. 4.3a evinces a spatially extended depression around the location of the preceding spectrum. We attribute the blob-shaped protrusions on the surface to remaining molecular parts (e.g., phenyl rings), thus considering molecular fragmentation (C–C bond scission) and the uptake of fragments via the STM tip.

Next, we inspected the monomers' response toward voltage pulses (3 V) of different pulse duration, which were applied under OFL conditions at pristine spots of a molecular island (Fig. 4.3b). As detected by subsequent STM imaging, depression-

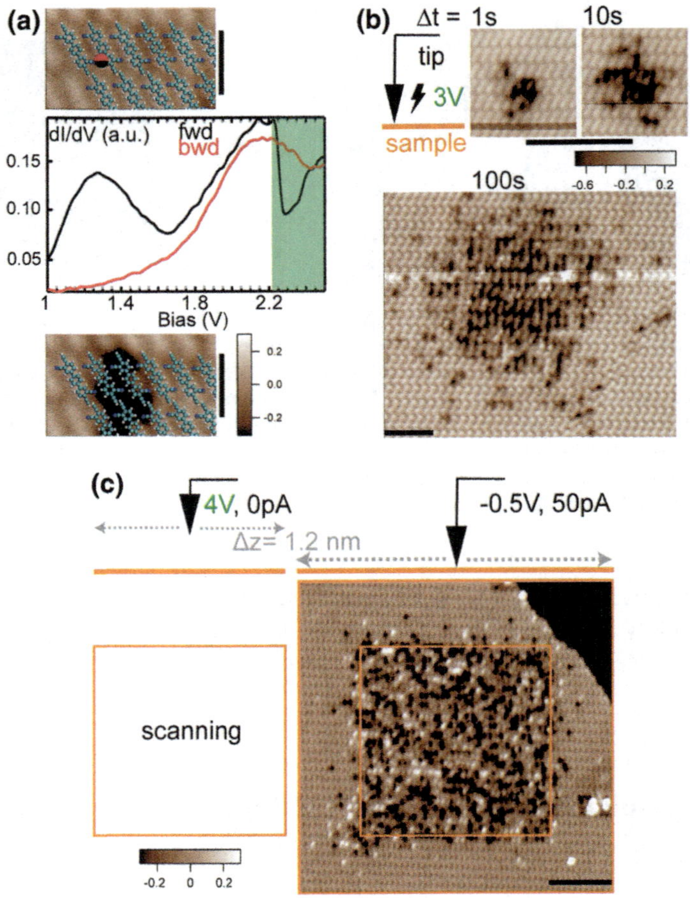

Fig. 4.3 Tip-induced monomer desorption. **a** STS spectrum on monomer (red-black circle) shows abrupt decline of *dI/dV* signal for $V \geq 2.2$ V (black curve). Superimposed molecular models with cyan C, blue N and white H atoms. Subsequent STM image reveals molecular desorption. **b** STM tip-induced desorption via voltage pulses with a duration of 1, 10 and 100 s. Subsequent STM topographs reveal spatial extension of tip-induced pattern. **c** High-voltage scanning of a square-shaped region after tip retraction from tunneling regime ($\Delta z = 1.2$ nm). Subsequent low-voltage STM image with approached tip (tunneling regime) of a larger region reveals square-shaped pattern within monomer island. Tunneling parameters V_t, I_t: **a** 1002 mV, 0.1 (top) and 0.05 nA (bottom); (b top) 2000 mV, 0.08 nA; (b bottom) −5 mV, 50 pA; 1000 mV, 0.08 nA; **c** −500 mV, 0.05 nA. Scale bars: **a** 2 nm; (b top) 4 nm; (b bottom) 8 nm; **c** 10 nm. STS setpoint parameters: 1002.5 mV, 0.05 nA, $V_{mod} = 7$ mV rms, $f = 733.47$ Hz, $T = 4.5$ K

like motifs are encountered within circular-shaped areas, whose diameters are directly proportional to the duration of the applied voltage pulse (1, 10 and 100 s). While the isotropic pattern may suggest a symmetric geometry of the STM tip, the depressions within the exposed region may be due to rotated CN-phenyl units as well as displaced and fragmented CN-DETP monomers.

The process might be triggered by electron-induced excitations during (resonant) tunneling or by the electric field within the tunneling junction. In order to quench processes associated with electron tunneling, the tip located on a pristine island without defects, is retracted by 1.2 nm. Then, a square region of $25 \times 25\,nm^2$ is scanned at 3 V with a scanning speed of 781.25 Å per second (196.8 s for the entire image) under this condition (no tunneling current). After approaching the tip and rescanning a larger region in the constant-current mode (−0.5 V, 50 pA), we encounter depression-like motifs within the square-shaped region, which was exposed to high-voltage scanning before (yellow contour in Fig. 4.3c). Since this pattern was generated by scanning without electron transfer between tip and sample, the depressions emanate from electric field-induced processes. We postulate that the electronegative CN side groups are responsible for the monomers' response to elevated tip voltages. Interestingly, the CN-DETP molecular adlayer can be used as a template to create extended patterns with the sub-nanometer precision of the scanning tip. Unlike photolithography techniques, the "writing" process within this methodology uses the electric field of a STM tip. The presence of bright blobs within the square-shaped pattern is assigned to remaining molecular fragments displaced from their pristine locations. Therefore, we assume that the depressions within the STM images in Fig. 4.3a–c are caused by tip-induced desorption from the surface and monomer adsorption on the tip. The imaging quality of the tip can be judged from the appearance of the pristine monomers within the island region outside of the exposed area in Fig. 4.3c, and smooth tip-sample contacts on the pristine Ag(111) surface can be used to obtain reproducible imaging conditions. From the apparent depressions within the exposed region, we conclude a weak molecule-substrate binding for CN-DETP/Ag(111). The latter needs to be considered for the heat-induced activation of terminal alkyne homocoupling.

4.4 From Monomers to Covalent Oligomers

After confirming the structural integrity of surface-adsorbed CN-DETP monomers, terminal alkyne homocoupling is initiated by annealing a sample with a molecular coverage of ≤ 1 ML at 350 K for 10–20 min. Resuming STM measurements reveal a significant decrease in molecular coverage after sample annealing, hence implying substantial desorption losses. We rationalize the latter on the basis of weak monomer-surface interactions, consistent with the tip-induced monomer desorption stated before (cf. Fig. 4.3). The STM topograph in Fig. 4.4a shows the dissolution of monomer islands toward the formation of chain-like dimers and oligomers. In addition to straight dimers, a dimer species with a kink motif is encountered at the edges of dense-packed islands. The STM image in Fig. 4.4b displays an aggregate of four Γ-shaped dimers whose orientation is concluded from their CN-related depressions. From the very good agreement with the superimposed gas-phase model, an enyne motif enclosing an angle of 83° is suggested.

In total, we encounter twelve orientations of non-straight dimer motifs on Ag(111), which are marked by color-coded Γ-shaped outlines referring to the corners of three

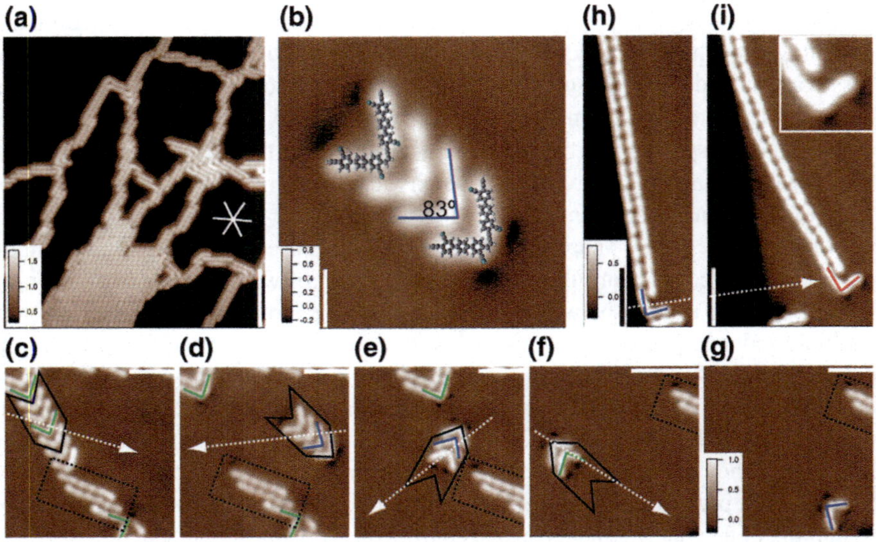

Fig. 4.4 Heat-induced coupling reactions at 350 K. **a** Dissolution of monomer phase and emergence of covalent dimers and oligomers. **b** Aggregate of four Γ-shaped dimers enclosing a bending angle of 83°. Superimposed dimer model for enyne species with CN units pointing to STM depressions (C: grey, N: cyan, H: white). **c–g** Tip-guided displacements of assembled dimers (within black outline) along indicated arrows reflect discrete orientations of Γ-shaped dimers (green, blue and red outline). Dashed rectangle represents a static reference. **h** Γ-shaped dimer motif (blue) terminating one covalent nanowire (of double-strand assembly). (i) Bent polymer nanowire after tip-induced manipulation on Γ-shaped kink along depicted arrow (red outline and inset). Tunneling parameters V_t, I_t: **a** 500 mV, 0.1 nA; **b–i** −100 mV, 0.1 nA. Scale bars: **a** 10 nm; **b** 1.6 nm; **c–e** 4 nm; **f** 6 nm; **g** 4 nm; **h, i** 4 nm

rectangles (green, blue and red) in the left part of Fig. 4.8c. In order to examine the influence of the Ag(111) surface on their orientations, lateral STM manipulation experiments are performed on assembled aggregates of these dimers. Figure 4.4c–g shows a series of consecutive STM images that were obtained after guiding the STM tip with the OFL parameters $V = -100\,\text{mV}$, $I = 100\,\text{nA}$ along the depicted arrows. The dashed rectangle serves as a static reference during the lateral manipulations on the aggregate within the black outline. After isolating a bunch of four Γ-shaped dimers in Fig. 4.4d, the following STM images reveal discrete rotations of the dimers upon tip-induced displacements of the aggregate. Hence, we consider selective dimer-surface interactions on Ag(111). The comparison of STM images before and after each manipulation step shows the decreasing number of dimers within the black outline, indicative of the occasional uptake of a dimer via the STM tip. In the final image (Fig. 4.4g), one isolated Γ-shaped dimer remains with four CN-related depressions according to the gas-phase model in Fig. 4.4b. Since the manipulation experiments indicate a rather weak dimer-substrate bonding, we do not attribute the non-linear motif to an organometallic silver–carbon intermediate, but explain it as the product of a side reaction proceeding without H desorption, but H transfer to trigger a single ethynyl → vinyl conversion (cf. Fig. 4.1b). This byproduct species formed through vinyl–ethynyl linkage coexists with the targeted linear ethynyl–ethynyl linkage emanating from homocoupling.

Extended polymer chains are observed at step edges, presumably because of their straight morphology and enhanced catalytical activity (undercoordinated Ag atoms) compared to flat Ag(111) terraces. Figure 4.4h shows two linear chains of interlinked CN-DETP molecules, one of them terminated by a byproduct motif (red outline and inset). The STM image after a lateral STM tip manipulation along the depicted arrow ($I = 100$ nA in OFL mode) indicates the collective displacement of all chain elements (Fig. 4.4i). Within in the isolated double-strand polymer, only the Γ-shaped kink exhibits unbound CN species (depressions). As shown in the inset, both molecules express cis isomerization, in contrast to the trans states in Fig. 4.4b, g. From the absence of depressions along the polymer backbone, it follows that all interlinked molecules are cis isomers engaging their CN groups in supramolecular interchain interactions, which represent a cohesive "zip" for the two polymer strands.

4.5 Extended Graphdiyne Wires and their Assemblies

The complete conversion from CN-DETP monomers to cyano-functionalized 3-2 graphdiyne (CN-GDY) wires on Ag(111) terraces is accomplished for annealing the sample at 400 K. As shown in the STM topograph in Fig. 4.5b, the only byproduct is an enyne compound, whose kink motif interrupts the linear growth (straight sections) of covalent chains. For the DETP precursor without CN side groups, similar preparation conditions ($T_{\text{deposition}} \leq 300$ K, $T_{\text{annealing}} = 400$ K for 10–20 min) on Ag(111) give rise to irregular coupling patterns with very short straight sections (Fig. 4.5a). Only a small yield of the targeted linear connection (green circles) was achieved due to the coexistence of bis-acetylide motifs (blue), non-linear twofold connections with varying bending angle (black) and threefold coupling nodes (red). This comparison impressively demonstrates how the CN side groups of CN-DETP strongly increase the chemoselectivity for homocoupling compared to DETP; thus allowing high-quality fabrication of extended GDY nanostrands on Ag(111) (Fig. 4.5b). Related attempts toward the shielding of reactive alkyne groups were conducted by Gao et al. utilizing long alkyl chains as steric side groups [5]. Accordingly, we attribute the increased chemoselectivity for homocoupling to the C≡N side group protecting both sterically and electronically the C≡C end group from side reactions, without getting involved in reactions.

In addition to controlling intermolecular linkage, the diatomic CN species mediates the supramolecular organization toward double- and poly-strand assemblies apparent in Fig. 4.5c. Figure 4.5e presents a high-resolution STM image of a double-strand polymer composed of two CN-GDY wires, each displaying an alternating sequence of C(sp^2) units (cyan outlines indicating molecular backbone) and C(sp) sections (orange outlines) (cf. Fig. 1.5a). From the absence of CN-related depressions outside of the double-strand assembly, we assume that all enantiomers express cis

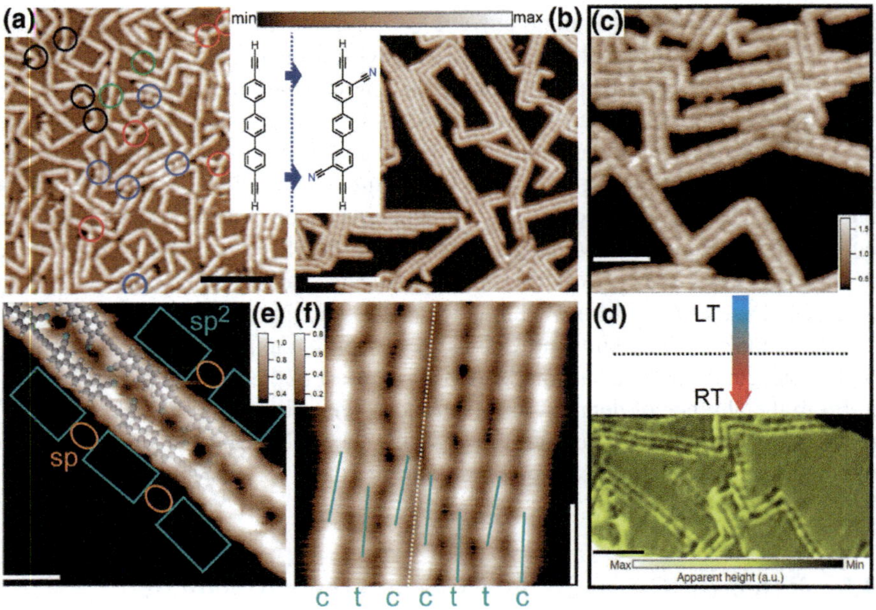

Fig. 4.5 Assemblies of GDY nanowires after heat-induced polymerization at 400 K. **a**, **b** Comparing homocoupling efficiency for DETP molecules (**a**) and CN-DETP molecules (**b**) on Ag(111). Byproducts in (**a**) are marked by color-coded circles. LT-STM (**c**) and RT-STM (**d**) results reveal complete conversion from CN-DETP monomers to assemblies of GDY polymer strands. **e** Double-strand assembly of GDY nanowires, whereby superimposed cis isomers indicate sequence of C(sp) and C(sp^2) units (cyan and orange outlines). **f** Poly-strand assembly containing enantiopure trans (t) and cis (c) nanowires according to indiciated monomers (cyan lines). CN groups point away from the dashed line. Tunneling parameters V_t, I_t: **a** −300 mV, 0.2 nA; **b** −10 mV, 0.1 nA; **c** 1000 mV, 0.1 nA; **d** −1000 mV, 0.07 nA; **e** 20 mV, 0.1 nA; **f** −100 mV, 0.05 nA. Scale bars: **a** 5 nm; **b** 10 nm; **c** 4 nm; **d** 5 nm; **e** 1 nm; **f** 2 nm

isomerization (superimposed molecular models), by which they impart CN-mediated interchain attractions between assembled polymer strands. The self-complementary character of the electrostatic interactions between the two covalent strands is reminiscent of DNA base pairing. Interestingly, the linear CN-GDY wires are displaced relative to each other in such a way, that each sp^2-hybridized monomer backbone (cyan outline) faces a sp-hybridized diyne unit (orange outline). Compared to single polymer strands, the double strand wires are associated with a higher mechanical stability and rigidity due to lateral attractions emanating from zip-like interchain bonding. In detail, the self-complementary arrangement of antiparallel CN$\cdots$phenyl bonds is supposed to favor the linear growth mode of the polymer strands. For the CN-mediated monomer assembly of the related monomer without terminal alkynes [6], supporting DFT modeling revealed a binding energy (BE) of ~0.1 eV for each CN-group [7]. Accordingly, the zipper-like double-strand assemblies of covalent strands should reveal a remarkably high thermal stability. Indeed, non-cryogenic

STM demonstrates stable imaging of these double-strand assemblies at room temperature (Fig. 4.5d).

Besides the predominance of double-strand wires, preparations with sufficiently high molecular coverage also lead to poly-strand assemblies, i.e., the stacking of several covalent molecular wires. Comparing their relative displacement along the polymer backbone (cyan lines indicate molecules) within each of the seven covalent polymer strands in Fig. 4.5f, one recognizes that the two wires next to the dashed line are not bound to each other through self-complementary CN$\cdots$phenyl bonding. Therefore, we assume that their CN-phenyl groups point away from the dashed line and label them as cis (c) polymers like the outer CN-GDY wires. From the regular displacement of the cyan lines within the remaining wires (molecular backbone facing diyne unit; cf. double-strand wire in Fig. 4.5e), we suppose that they express isochiral trans isomers along their polymer backbones (labeled by t (trans)), in order to maximize interchain interactions on both sides of the chains. The results demonstrate that polymers composed of isomeric building blocks can exhibit remarkable supramolecular recognition and adaptation phenomena. For comparison, we mention the coupling reactions of CN-DETP on Ag(110) resulting in enantiomorphic 2D domains of stacked trans Ag-bis-acetylide nanowires (cf. Fig. 3.3d–f).

4.6 Mechanical Flexibility of CN-GDY Molecular Wires

In the context of molecular-based engineering, polymeric building blocks need to exhibit high mechanical flexibility and stability in order to endure different processing steps, e.g., the transfer of polymer scaffolds on top of insulating platforms or contacting with metallic electrodes. Accordingly, we utilize STM tip manipulations to inspect the structural integrity, bending flexibility and mechanical stability of isolated CN-GDY nanowires. Two STM manipulation techniques, i.e., vertical and lateral manipulation, are utilized to isolate them from their assemblies. Figure 4.6a–c exemplifies the former technique with the STM tip placed on the end of a double-strand polymer (black/red marker), as shown in the upper STM image. For the $I(\Delta z)$ spectrum acquired in the OFL mode, the tip was approached toward the surface (forward (fwd) spectrum) and subsequently retracted (backward (bwd) spectrum) while recording the current as a function of varying tip-sample distance. The fwd spectrum shows an exponential behavior of the tunneling currect with respect to the tip-sample separation until reaching the saturation regime of the current preamplifier ($I \geq 10\,\text{nA}$). Since the bond formation between STM tip and termination of the molecular chain occurred in this regime, the associated jump in the current signal is not visible within the fwd spectrum during this vertical tip approach. The subsequent bwd spectrum starting from the saturation regime (red plateau) shows a sudden decline in the $I(\Delta z)$ trace at $\Delta z = -0.28\,\text{Å}$ (close to the initial tip height denoted as $\Delta z = 0$). We assign the abrupt decrease in current to the detachment of the molecular wire from the tip. After the vertical manipulation, the subsequent STM image in Fig. 4.6b reveals the rearrangement of several isolated GDY wires recognized by their CN-related depressions along the polymer backbone. From the location of surface impurities (yellow arrow), we recognize the detachment of a nanowire from a step edge. Besides the disappearance of several strands indicative of polymer uptake via the tip, the structural integrity of the remaining nanochains hints to a remarkable mechanical stability.

Magnifying the region within the yellow outline (Fig. 4.6c), a double-strand assembly with some trans conformers is recognized after the manipulation. As expected, the CN moieties of superimposed molecular models coincide with depressions in the STM contrast. When assigning them with the states "0" ("1") for CN-phenyl groups pointing to the left (right) side, each strand can be translated into a binary code. The adjoined polymer model obtained by mirroring the superimposed model is associated with the inversion of the original binary code. With each polymer strand representing a binary block, one could use an isolated CN-functionalized GDY nanochain to establish binary encoding at the molecular level.

For this objective, lateral STM manipulations are used as a more controlled way to isolate polymer wires from CN-mediated interactions with other molecular wires or step edges. To isolate a linear molecular wire from a step edge, the tip is approached on the pristine Ag(111) surface to a tip-sample distance corresponding to $V_t = 100\,\text{mV}$ and $I_t = 75\,\text{nA}$ within the OFL mode. Subsequently, the tip is slowly guided along the depicted arrow (purple) passing through the step edge and termination of the molecular wire (Fig. 4.6d). Due to electrostatic attractions between the tip and molecular wire, the latter is forced to follow the moving tip to the targeted position via a pushing and pulling movement [8]. As a consequence of the lateral manipulation, the subsequent STM image reveals the detachment of a polymer section from a step edge without disrupting on-surface formed bonds between CN-DETP monomers. As indicated by the depressions along the polymer backbone, CN moieties are pointing toward the step edge.

Figure 4.6e–h reveals that also double-strand assemblies aligned along step edges can be separated via lateral manipulation. To this end, the STM tip is placed on a chain termination (terminal alkyne) with a setpoint of $V_t = -100\,\text{mV}$ and $I_t = 0.1\,\text{nA}$. Then, the current I_t is increased in feedback mode (typically to $\sim$30 nA) until noticing a sudden jump in the z-signal, attributed to tip-polymer bonding. By slowly moving the tip with a speed of 1 nm/s along the purple line, one polymer strand gets removed. Since the CN moieties of the remaining wire in Fig. 4.6f point away from the step edge, we conclude that interchain binding is preferred over binding to step edges. After removing the aggregate of short oligomer chains within the white outline (purple arrow with number 1.), the targeted polymer wire can be isolated in a stepwise fashion (steps 2.–4. indicate chronological order). Each manipulation step along the purple vector liberates one chain section from the step edge and moves it to a new position, where the tip is retracted from the surface-adsorbed polymer. The STM image after these manipulation steps (Fig. 4.6g) evinces a strongly bent arrangement of the CN-GDY strand, i.e., a remarkably small bending radius of 19 Å (circle). Interestingly, a further manipulation step via the STM tip (displacement along purple arrow) transforms this artifically created arrangement back to its linear configuration

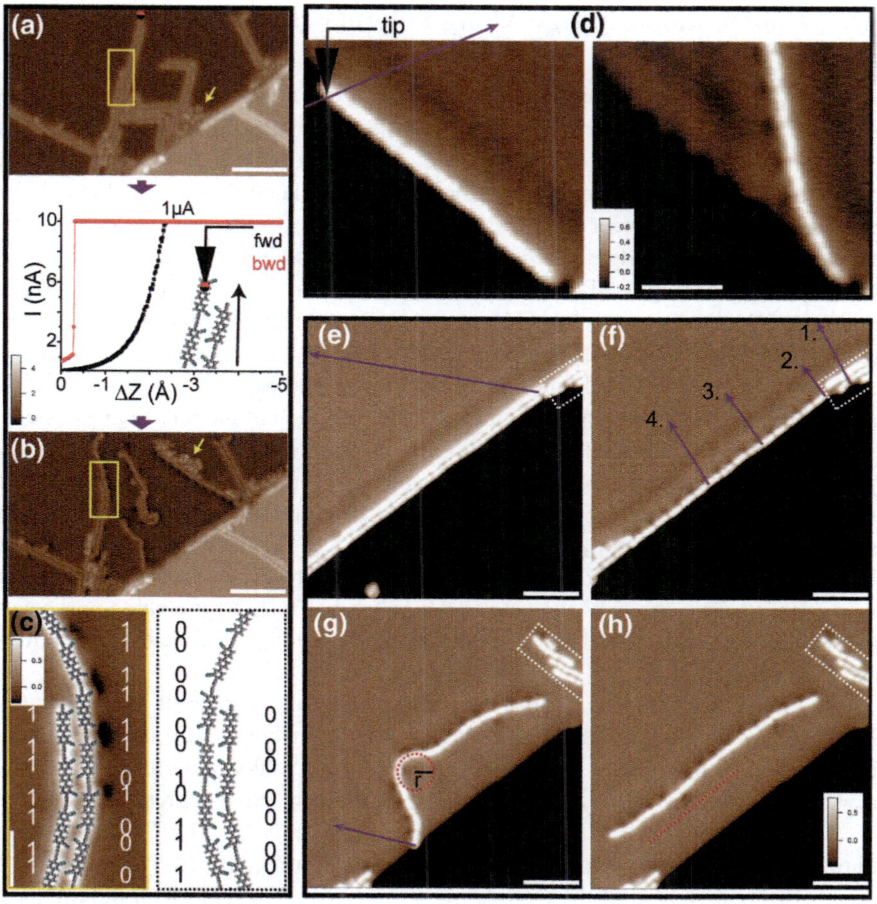

Fig. 4.6 Isolation of single-strand polymer via vertical (**a**–**c**) and lateral (**d**–**h**) tip manipulation. **a** Forward (black) and backward (red) I(z) spectrum at termination of double-strand polymer (red/black marker). **b** Subsequent STM image reveals single-strand polymers with CN-related depressions. **c** High-resolution STM topograph shows trans isomers within yellow outline of (**b**). The states 0 (1) define left-pointing (right-pointing) CN groups. Adjoined molecular model depicts mirror structure and associated binary code. **d** Isolation of a single-strand wire from a step edge via lateral manipulation. **e** Double-strand wire at a step edge before performing single-polymer manipulation along purple line. **f** After the removal of one covalent strand, short oligomers within rectangle (1.) and remaining molecular wire are laterally displaced in a stepwise fashion (2.–4.). **g** Extreme bending flexibility according to circle with bending radius of 19 Å. **h** Subsequent straightening of the molecular wire. Tunneling parameters V_t, I_t: **a** 1000 mV, 0.1 nA; **b, c** −100 mV, 0.1 nA; **d**–**f** 100 mV, 0.1 nA; **g, h** −100 mV, 0.1 nA. Scale bars: **a, b** 10 nm; **c** 2 nm; **d** 4 nm; **e**–**h** 6 nm. STS setpoint parameters for I(z) sweep: −100 mV, 0.1 nA, $\Delta z = 500$ pm, $T = 4.5$ K

(Fig. 4.6h). The astonishing flexibility and structural integrity of the CN-GDY wire during this reversible procedure is ascribed to the mechanical strength and bending flexibility of sp-hybridized chain elements (diyne units) [9].

4.7 Information Storage Along Polymer Backbone

After developing reliable protocols to isolate CN-GDY nanowires, their capability toward information storage is scrutinized. The schematic illustration in Fig. 4.7a proposes a mechanism for conformational switching of CN-phenyl groups via the STM tip. The latter are recognized from their CN-related depressions, which are clearly imaged at a setpoint of $V_t = \pm 100\,\mathrm{mV}$ and $I_t = 0.1\,\mathrm{nA}$. For single molecular wires attached to step edges, we firstly released a section of the molecular wire from the step edge (cf. Fig. 4.6d). First, we placed the tip at a setpoint of $V_t = 100\,\mathrm{mV}$ and $I_t = 0.1\,\mathrm{nA}$ on pristine Ag(111) next to a CN-induced depression, switched off the feedback loop and lowered the tip by 2 Å (thereby increasing I_t). Next, the tip was guided at a speed of 1 nm/s along the orange vector in Fig. 4.7b, in order to push the CN group to the opposite side via a 180° rotation of the relevant CN-phenyl group. To avoid the displacement of the entire nanowire on Ag(111), one part of the polymer strand remains bound to the step edge (cf. Fig. 4.6d). As evident from the subsequent STM image, the pertaining CN-related depression appears at the opposite side with respect to the polymer backbone (Fig. 4.7c). The switching procedure can be repeated on individual CN groups (Fig. 4.7c, d), or get reversed by starting the manipulation from the opposite side of the polymer. Thus, we are able to associate the eight CN-phenyl units of this polymer section with eight bits (one byte) for binary information storage.

For single molecular wires that were completely removed from step edges (cf. Fig. 4.6h), we applied a slightly different routine to enable conformational switching, in order to avoid displacement of the entire molecular chain during manipulation. To this end, the tip is initially centered on the respective depression (setpoint: $V_t = -100\,\mathrm{mV}$ and $I_t = 0.1\,\mathrm{nA}$) and approached by ~3 Å under OFL conditions until a sudden jump in I_t occurred (several hundred nA), which is attributed to a conformational change of the CN-phenyl group forming a bond with the tip apex. By retracting the tip by 1.5 Å, we observe a hysteresis in $I(z)$ (not shown), which indicates tip-polymer bonding. To perform a switching event, the tip is then laterally moved perpendicular to the nanowire with a speed of 1 nm/s (cf. Fig. 4.7b). After reaching the opposite site of the polymer backbone, the feedback loop is closed to restore the flat-lying configuration of the CN-phenyl unit after successful 180° rotation.

Following this method, a pentameric section of an isolated CN-DETP polymer strand is used to consecutively write the letters "T", "U", "M" (acronym for "Technical University Munich") in binary code (ASCII values), as depicted on the left side of each letter. According to Fig. 4.6c, the states "0" and "1" are defined as left- and right-pointing CN units, respectively. Since each STM image section in Fig. 4.7e represents one ASCII letter, the related storage density amounts to ~0.34 bit/nm^2. This nominal value corresponding to a density of 219 terabit per square inch follows from the presence of two CN-phenyl units for each isomer along the covalent strand, and from assuming a width of ~3 nm perpendicular to the wire. For comparison, this density is approximately two orders of magnitude larger than the storage density

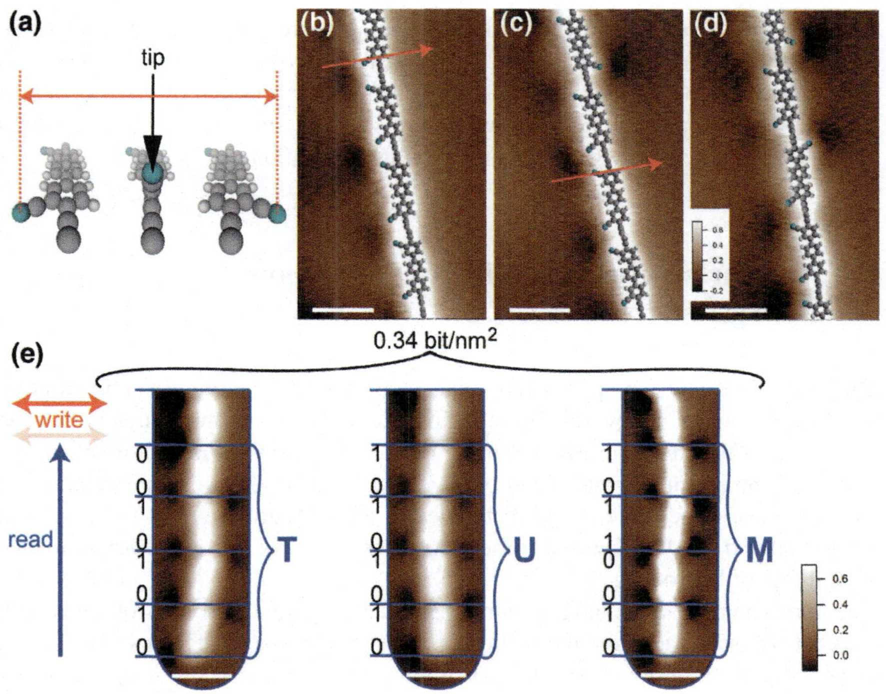

Fig. 4.7 **a** Schematic illustration for tip-induced switching of isomer states. Orange lines in (**a–c**) indicate trajectory of STM tip during switching. **b–d** STM images show sequence of two switching events, indicated by new positions of respective CN-related depressions. **e** Binary encoding and information storage of the acronym "TUM". Tunneling parameters V_t, I_t: **b** 100 mV, 0.1 nA; **c–e** −100 mV, 0.1 nA. Scale bars: **b–d** 1.4 nm; **e** 2 nm

in recent hard drives. A further increase in storage density might be accomplished by tuning the employed precursor design, i.e., omitting the central phenyl ring of CN-DETP in the molecular synthesis route.

To estimate the thermal stability of this conformational switch, we mention a recent calculation on the cis ↔ trans isomerization barrier of the related molecular derivative without terminal alkynes. For each monomer confined within a molecular layer on Ag(111), the barrier amounts at least to 0.49 eV [10]. This barrier is surpassed by the 180° rotation of a CN-phenyl unit corresponding to one switching event. Based on recent calculations on the thermal stability limits of high-density data storage systems, it was shown that an uniaxial barrier of 40 $k_B T$ is suitable for technological applications [11]. The isomerization barrier should therefore guarantee the stability of the isomer state up to a critical temperature of 142 K. This value significantly exceeds the stability of antiferromagnetic few-atom structures, which become unstable already at 4 K [12]. In the context of CN-DETP, additional contributions to the isomerization barrier are expected due to an increased surface-molecule bonding stemming from butadiyne linkage groups (C≡C−C≡C) along the covalent strand, as compared to the single CN-DETP isomer. Therefore, we assume that the $2n$ CN-phenly units along the backbone of an isolated CN-GDY nanowire comprised of n CN-DETP molecules remain unchanged (no 180° rotations) for temperatures exceeding the proposed value. Accordingly, its 2^{2n} possible conformational states are supposed to be highly stable against thermal fluctuations.

4.8 On-Surface Synthesis of CN-GDY Nanowires on Ag(100)

This section presents an approach to enhance the linear growth mode toward more extended CN-GDY nanowires. From a closer look at the polymer wires within the large-scale STM image in Fig. 4.8a, one encounters twelve orientations of the Γ-shaped enyne species (byproducts) on Ag(111). Their orientations are depicted as the corners of three color-coded rectangles in the left part of Fig. 4.8c. Aiming at more extended CN-GDY strands (linear polymer sections), it is necessary to reduce the number of byproducts.

Lateral manipulations have shown their discrete orientations and rotations with respect to the substrate's symmetry directions (Γ-shaped kinks in Fig. 4.4). Hence, we suspect that the number of orientations of the enyne compound (byproducts) is dictated by the symmetry of the surface, and propose the use of a less symmetric surface in order to reduce the number of possible orientations for the polymer chains. While the anisotropic Ag(110) surface triggers the organometallic linkage of CN-DETP (cf. Chap. 3), the fourfold symmetric Ag(100) surface also steers alkyne homocoupling. The STM topograph in Fig. 4.8b displays CN-GDY nanowires created on Ag(100) under similar conditions as on Ag(111) (annealing at 400 K for 15 min). Indeed, more extended straight polymer sections and less byproducts are encountered on Ag(100) compared to Ag(111). Their orientations on Ag(100) represent the corners of two (instead of three) color-coded rectangles (right part of Fig. 4.8c).

By counting the amount of Γ-shaped dimer motifs in Fig. 4.8a, b within the CN-GDY strands grown on Ag(111) and Ag(100), and dividing the latter by the total number of molecules within the STM images, the percentage amount of byproducts is obtained. The color-coded bars of the histogram in Fig. 4.8c subdivide this amount according to their orientations (grouped by three (two) squares for Ag(111) (Ag(100)). Within the statistical error, 17% of all monomers on Ag(111) converted into byproducts, while on Ag(100), only 8% are involved in the pertaining side reaction. Interestingly, on Ag(100), only a few Γ-shaped motifs are oriented according to the corners of the red rectangle in Fig. 4.8c, while on Ag(111), the histogram reveals a nearly equal distribution of all three subgroups (corners of three rectangles). The reduced number of byproducts on Ag(100) should reinforce the linear growth mode toward more extended GDY-like molecuar wires.

The chemoselectivity toward the targeted linear ethynyl–ethynyl coupling motif was compared for both surfaces by calculating the reaction yield R for homocou-

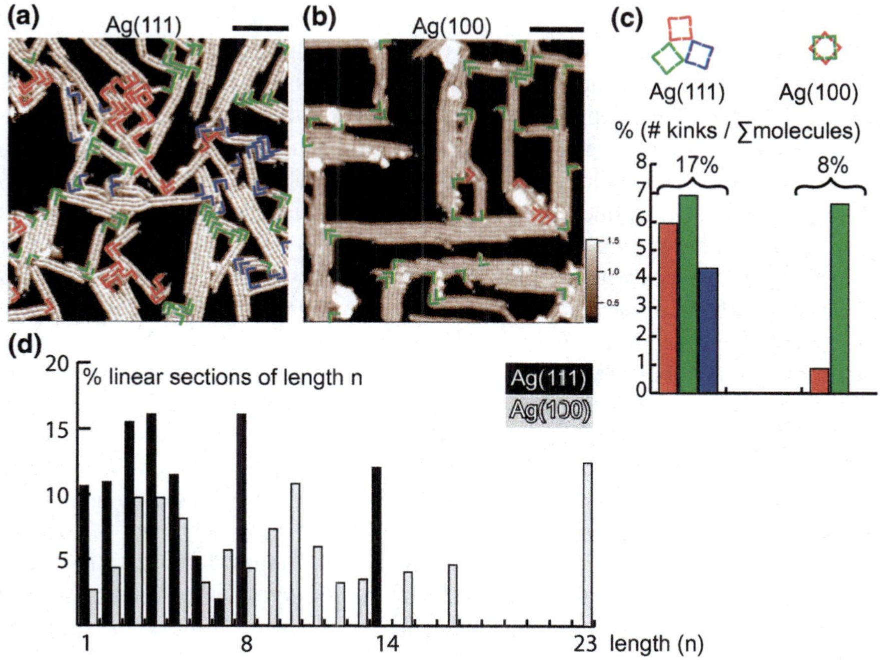

Fig. 4.8 Comparing on-surface coupling on Ag(111) and Ag(100). **a**, **b** Large-scale STM topographs on Ag(111) (**a**) and Ag(100) (**b**) for statistical evaluation. Byproducts are marked by color-coded Γ-shaped outlines. **c** Their orientations on Ag(111) (Ag(100)) coincide with the corners of three (two) squares. Histogram with color-coding referring to kink orientations reveals the percentage amount of byproducts within STM images. **d** Length distribution of CN-GDY strands for Ag(111) (black bars) and Ag(100) (grey bars), i.e., percentage share of linear polymer sections of length n (in monomers). Chain terminations are marked by asterisks in (**a**, **b**). Tunneling parameters V_t, I_t: **a** 1000 mV, 0.1 nA; **b** 100 mV, 0.1 nA. Scale bars: **a**, **b** 10 nm

pling in the following way (see Ref. [13]). In the first step, we determined the total number of molecules within each STM image. Since two terminal alkyne groups are engaged in each homocoupling motif, the number of butadiyne bridges corresponds to the number of molecules participating in the targeted reaction. After substracting this number from the total amount of molecules, the remaining quantity is composed of non-linear byproducts (Γ-shaped kinks) and terminal alkynes not involved in intermolecular linkage (asterisks indicate chain terminations) [13]. By counting the amount of vinyl–ethynyl catenation motifs (byproducts) and chain terminations within the STM topographs for Ag(111) (Ag(100)) in Fig. 4.8a, b, we found out that 617 (491) of 824 (572) molecules are engaged in homocoupling. The reaction yield for homocoupling R, defined by the ratio of the latter, accounts for 75% on Ag(111) and 86% on Ag(100). Thus, we conclude that the chemoselectivity for on-surface homocoupling is only slightly higher on Ag(100) compared to Ag(111), which is due to a high concentration of chain terminations (asterisks) on Ag(100) despite the presence of less byproducts. The blob-shaped impurity clusters in the STM image of Fig. 4.8b might conceal further butadiyne bridges related to homocoupling.

A further important parameter for comparison is the length distribution of CN-GDY strands, which represent the linear sections of the covalent strands. According to Fig. 4.8a, b, their lengths are delimited by Γ-shaped kinks and chain terminations. The abscissa of the diagram in Fig. 4.8d represents the length n (in monomers) of CN-GDY strands, i.e., linear polymer sections without including the two molecules of each vinyl–ethynyl linkage. Its vertical axis displays in percentage values how many of the molecules within all linear strands are contained within linear sections of length n, which are terminated via side reactions and/or by unreacted terminal alkynes. On the basis of this analysis, the majority of molecules on Ag(111) (black bars) is incorporated in straight polymer sections with lengths $n \leq 8$ monomers. On Ag(100) (grey bars), the majority of CN-GDY strands exhibits a length of $n \leq 17$ monomers, while the maximum length corresponds to $n = 23$. To obtain more statistics for $n \geq 14$, it would be necessary to evaluate more STM topographs. Comparing the two distributions, one notes that CN-GDY nanowires synthesized on Ag(100) are significantly more extended, which we explain by a lower amount of byproducts exhibiting less orientations than on Ag(111) (six compared to twelve Γ-shaped motifs in the STM image of Fig. 4.8b). Since the less-symmetric Ag(100) surface exhibits less orientations for polymer nanowires as compared with Ag(111), the former can be used to create more extended CN-GDY strands facilitating more regular poly-strand assemblies.

4.9 Conclusion

The monomer assembly and on-surface reaction of the novel precursor CN-DETP was studied on the Ag(111) surface. A supramolecular tecton involving CN groups and terminal alkynes in a self-complementary fashion triggers the formation of a dense-packed arrangement of the monomer phase. The monomer's sensitivity to elevated tip voltages is rationalized by electric field-induced molecular desorption due to a weak surface-monomer interaction. Heat-induced homocoupling reactions at 350 K give rise to short oligomers at the edges of monomer islands and covalent polymer strands aligned along step edges. Compared to heat-induced on-surface reactions with the DETP linker, the additional CN side groups of CN-DETP lead to significantly higher reaction yields toward the targeted ethynyl–ethynyl linkage group (homocoupling). The only byproduct is an enyne motif expressing twelve discrete orientations on Ag(111), which change upon tip-induced lateral displacements. After sample annealing at 400 K, all molecules participate in polymerization reactions toward covalent molecular chains, whose straight sections represent GDY-like nanowires. The CN groups not only promote the linear growth mode toward CN-GDY nanowires, but also mediate their regular assemblies via electrostatic interchain attractions. Double-strand chains based on self-complementary CN$\cdots$phenyl bonding between cis enantiomers coexist with poly-strand assemblies incorporating covalent chains of trans isomers. The CN-related zip-like interchain bonding within these 1D molecular wires leads to high thermal stability, as evicend by STM measurements at room temperature.

Both STM-based vertical and lateral manipulations can be used to isolate single covalent strands from their assemblies. Due to their sp-hybridized butadiyne sections, the isolated chain reveals high mechanical strength and flexibility, thus allowing reversible transformations between different bending conformations. Thanks to the expression of CN-related depressions in the STM contrast, the isomer states of all molecules along the polymer backbone are visible. Encouraged by the presence of a substantial cis $\leftrightarrow$ trans isomerization barrier indicative of a high thermal stability for each isomer state (above $\sim$140 K), we explored the potential of isolated CN-functionalized nanowires toward information storage at the nanoscale. For the single-molecule manipulations, the STM tip is utilized to flip single CN-phenyl moieties along the polymer backbone, as demonstrated by writing the acronynm "TUM" in binary code.

Terminal alkyne homocoupling of CN-DETP is also accomplished on Ag(100), where significantly less byproducts and more extended CN-GDY molecular wires are formed, as compared to Ag(111).

By using a vicinal surface with closely spaced step edges, the amound of byproducts may get further reduced, and the balance between double- and poly-strand CN-GDY assemblies might get controlled through the distance between steps.

Our methodology of combining on-surface alkyne coupling with CN-mediated supramolecular recognition opens up diverse variation possibilities toward the fabrication of well-defined nanomaterials made up of C(sp^2)-based building blocks connected through linear butadiyne bridges. Due to the mechanical flexibility, thermal stability and switching behavior of CN-GDY nanowires, they are highly suited for new applications in the field of molecular electronics. Furthermore, the CN groups along the covalent wires may represent n-doping centers with high potential for flexible bandgap tuning.

References

1. Cirera B, Zhang Y-Q, Björk J, Klyatskaya S, Chen Z, Ruben M, Barth JV, Klappenberger F (2014) Synthesis of extended graphdiyne wires by vicinal surface templating. Nano Lett 14:1891–1897
2. Björk J, Zhang Y-Q, Klappenberger F, Barth JV, Stafström S (2014) Unraveling the mechanism of the covalent coupling between terminal alkynes on a noble metal. J Phys Chem C 118:3181–3187
3. HyperChem^{TM} Release 7.0 (MolecularModeling), Hypercube, Inc. http://www.hyper.com/
4. Arras E, Seitsonen AP, Klappenberger F, Barth JV (2012) Nature of the attractive interaction between proton acceptors and organic ring systems. Phys Chem Chem Phys 14:15995–16001
5. Gao H-Y, Wagner H, Zhong D, Franke J-H, Studer A, Fuchs H (2013) Glaser coupling at metal surfaces. Angew Chem Int Ed 52:4024–4028
6. Marschall M, Reichert J, Seufert K, Auwärter W, Klappenberger F, Weber-Bargioni A, Klyatskaya S, Zoppellaro G, Nefedov A, Strunskus T, Wöll C, Ruben M, Barth JV (2010)

Supramolecular organization and chiral resolution of p-terphenyl-mdicarbonitrile on the Ag(111) surface. Chem Phys Chem 11:1446–1451
7. Abbasi-Pérez D, Recio JM, Kantorovich L (2014) Building motifs during self-assembly of para-terphenyl-meta-dicarbonitrile on a metal surface: a gas-phase study. J Phys Chem C 118:10358–10365
8. Hla H-W (2005) Scanning tunneling microscopy single atom/molecule manipulation and its application to nanoscience and technology. J Vac Sci Technol B 23:1351–1360
9. Szafert S, Gladysz J (2003) Carbon in one dimension: structural analysis of the higher conjugated polyynes. Chem Rev 103:4175–4206
10. Abbasi-Perez D, Manuel Recio J, Kantorovich L (2015) The role of isomerization in the kinetics of self-assembly: p-terphenyl-m-dicarbonitrile on the Ag(111) surface. Phys Chem Chem Phys 17:11182–11192
11. Charap SH, Lu P-L, He Y (1997) Thermal stability of recorded information at high densities. IEEE Trans Magn 33:978–983
12. Loth S, Baumann S, Lutz CP, Eigler DM, Heinrich AJ (2012) Bistability in atomicscale antiferromagnets. Science 335:196–199
13. Liu J, Chen Q, Xiao L, Shang J, Zhou X, Zhang Y, Wang Y, Shao X, Li J, Chen W, Xu GQ, Tang H, Zhao D, Wu K (2015) Lattice-directed formation of covalent and organometallic molecular wires by terminal alkynes on Ag surfaces. ACS Nano 9:6305–6314

Chapter 5
Metal Alkynyl π Complexes

This chapter is a modified version of the manuscript and supporting information created by the author of this thesis for publication in the journal "ACS Nano" (see chapter "List of publications").

5.1 Introduction

Aiming at the organization of magnetic nanoobjects, such as single atoms [1], small clusters [2], or single-molecule magnets [3], an interfacial adlayer can serve as a template to host and arrange the guest species in a highly selective fashion. One-atom thick sheets of graphene and boron nitride proved to be suitable templates for the trapping of metallic clusters [4–8] at specific sites of the substrate-related Moiré lattices. Open-porous metal-organic coordination networks (MOCN) [9] were employed for the selective decoration with transition metal clusters [10] and confinement of single atoms [11]. A self-assembled and highly ordered layer of diphenyl oxalic amide molecules was used to selectively bind monomeric cobalt on top of benzene rings, thereby establishing a well-ordered Co superlattice [12].

In this chapter, we employ an organic species with six internal alkynes reflecting multiple organometallic binding sites. Exposing an organic template of self-assembled HEB molecules on Ag(111) to a beam of Co atoms leads to the formation of mononuclear (C_1) and dinuclear compounds (C_2) expressing site-selective metal-alkynyl interactions. Combining high-resolution STM and state-of-the-art DFT modeling we elucidate the structure of the host-guest complexes and report on their adsorption behavior. Unlike the σ-like organometallic Ag-bis-acetylide chains presented in the third chapter, this π-type complex expresses bis(η^2-alkynyl) tweezer motifs [13, 14], each stabilizing a single Co atom (see Fig. 1.3c).

Such metal alkynyl complexes are attractive for various research fields such as nonlinear optics [15], intramolecular charge transfer systems [16] and photo-

R. Hellwig, *Alkyne-Based Nanostructures on Silver Substrates*, Springer Theses,
https://doi.org/10.1007/978-3-030-00997-7_5

voltaics [17] (see Sect. 1.3.4). In addition, they constitute the key building blocks for metallopolymers, an emerging class of functional soft materials [18]. Moreover, organometallic Ni-alkyne bonding facilitated the post-functionalization of an adaptable metal-organic network [19]. Recently, we reported the surface-guided formation of an original organocobalt complex from a precursor with one internal alkyne moiety (see Fig. 1.3e). The established organometallic compound holds no counterpart in solution chemistry, thus highlighting the intriguing phenomena triggered by the interfacial conditions [20]. Related organocobalt structures are highly relevant for chemical research, where organocobalt compounds represent indispensable ingredients for numerous synthesis pathways based on the Pauson–Khand and Nicholas reaction, and also the hydroformylation reaction [21].

5.2 Organic Layer

The sixfold symmetric alkyne derivative HEB (Fig. 5.1a) with stochiometric formula $C_{36}Si_6H_{54}$ was synthesized by the chemist Zhi Chen from the Ruben group [22]. The chemical structure is comprised of a substituted benzene compound connected to six internal alkynes, each terminated by a trimethylsilyl (TMS) group. Figure 5.1b–d shows different perspectives of a ball-and-stick molecular model established by geometry optimization via gas-phase DFT calculations. The calculated distances between adjacent alkynes are shown in the top view of the molecule (Fig. 5.1c), the alkyne's bending flexibility is indicated by the out-of-plane (in-plane) angle ε (φ) (Fig. 5.1b, c), and rotations of the TMS groups around σ bonds are indicated by the angle γ (Fig. 5.1d). The mutual orientations of the six TMS groups induce chirality within surface-adsorbed molecules.

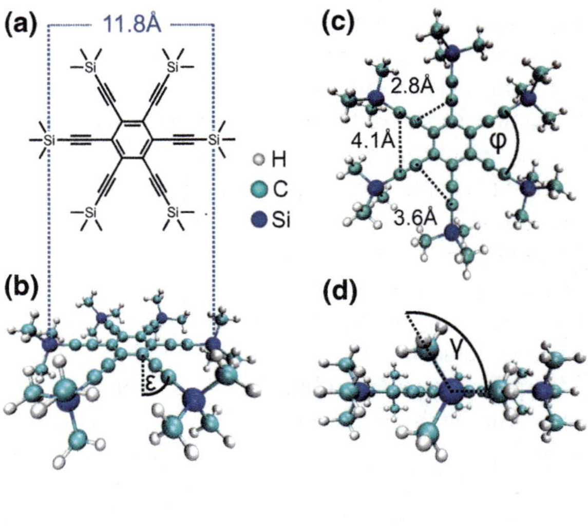

Fig. 5.1 **a** Chemical structure model of HEB. **b** Gas-phase DFT-calculated atomic ball-and-stick model for HEB. Color coding: carbon (cyan), silicon (blue) and hydrogen (white) atoms. **c**, **d** Top and side view of DFT-calculated model

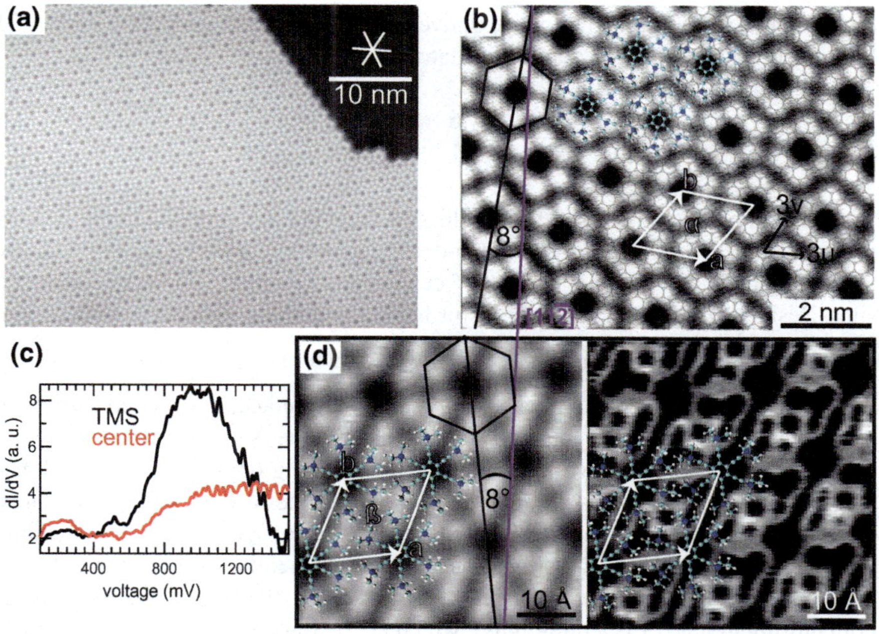

Fig. 5.2 Organic phase showing α domain (**b**) and β domain (**a**, **d**) according to depicted molecular unit cell vectors $\vec{a}$, $\vec{b}$ with respect to atomic unit cell vectors $\vec{u}$, $\vec{v}$. **a** Large-scale STM image of a HEB island; star depicts high-symmetry directions of Ag(111). **b** High-resolution image with superimposed unit cell, molecular models tentatively positioned on Ag(111) registry. Molecules of α and β domain are rotated by $\pm 8°$ (black line with outlined molecule) with respect to $[11\bar{2}]$ direction (purple line). **c** STS spectrum on top of a molecular lobe (TMS) reveals LUMO peak. **d** Corresponding *dI/dV* map at LUMO energy (right) is shown together with simultenously acquired constant-height topograph (left) (image resolution: 96 × 96 pixel^2). Tunneling parameters V_t, I_t: **a** −100 mV, 100 pA; **b** −10 mV, 49 pA; initial setpoint: **c** 100 mV, 75 pA, $V_{mod} = 5$ mV rms, $f =$ 933.5 Hz, $T = 4.5$ K; **d** 1000 mV, 1012 pA, $V_{mod} = 4$ mV rms, $f = 733.5$ Hz, $T = 4.4$ K

The powder of purified HEB molecules was thoroughly degassed and sublimated in UHV at a crucible temperature of 430 K. In a first step, the chemical vapor deposition of HEB on the clean Ag(111) surface, kept at 450 K, was performed in order to grow a sub-ML molecular interface. The high deposition temperature was chosen in order to avoid the adsorption of defect molecules within organic domains, i.e., a minute concentration of synthesis byproducts missing one TMS-alkyne group. After transferring the sample to a LT-STM, the measurements show extended, homogenous and largely defect-free organic islands (Fig. 5.2a), indicative of a high molecular mobility and weak intermolecular interactions allowing structural autocorrection mechanisms during self-assembly. The hexagonal symmetry and dense packing scheme of the islands is rationalized by the predominance of attractive van-der-Waals interactions between bulky TMS end-groups. When superimposing a Ag(111) lattice model and molecular models to high-resolution STM data (Fig. 5.2b), the six bright lobes surrounding a central void coincide with the six TMS-alkyne groups. Based on high-resolution STM imaging, we tentatively suggest a commensurate $\sqrt{31} \times \sqrt{31}$ R 8.9° adsorption structure reading in matrix notation,

$$\begin{pmatrix} \vec{a} \\ \vec{b} \end{pmatrix} = \begin{pmatrix} 6 & -1 \\ 1 & 5 \end{pmatrix} \begin{pmatrix} \vec{u} \\ \vec{v} \end{pmatrix},$$

whereby $\vec{u}$ and $\vec{v}$ are the primitive vectors of Ag(111). As justified in the following analysis of Co-HEB units within Co-decorated organic islands (see Fig. 5.3), we tentatively postulate that the molecules' centers adsorb on sixfold symmmetric top sites. Besides this domain labeled α, a further domain orientation exists with its unit cell reading in matrix notation: $\begin{pmatrix} 5 & 1 \\ -1 & 6 \end{pmatrix}$, as shown in Fig. 5.2d. The existence of this domain β is rationalized by a combination of the molecule's sixfold symmetry and alignment with respect to a crystallographic symmetry axis. A hexagonal contour enclosing a molecule with an associated line along opposite molecular lobes (Fig. 5.2b, d) reveals that the orientation of the molecules differs for both domains. We found a rotation of $\pm 8°$ with respect to the three mirror symmetry axes coinciding with the substrate's $\langle 11\bar{2} \rangle$ directions (purple line representing one of them). Therefore, the molecular arrangement is chiral and is expressed in two enantiomorphic domains (aggregate of isochiral units), which are distinguished by a 16° rotation of the unit cell.

In addition to the structural properties of the organic phase, its electronic properties are characterized by performing *dI/dV* spectroscopy. The STS profile in Fig. 5.2c reveals the existence of a LUMO resonance localized on the six molecular protrusions of HEB (bright blobs within STM contrast). The corresponding 2D *dI/dV* map (constant-height mode) at 1 V reflects the spatial delocalization of the identified LUMO signature (right part of Fig. 5.2d). According to the superimposed molecular models on top of the LUMO map, bright contours reflecting unoccupied DOS coincide with protruding methyl groups within the terminal groups of HEB. The latter are suspected to be more decoupled from the surface (higher physical z-heights) than the deeper lying alkynes and arene ring of HEB, which appear as depressions within the *dI/dV* map. Interestingly, the *dI/dV* map reveals two alternating appearances within the six TMS units of assembled HEB i.e., three TMS units appear more dark (smaller unoccupied DOS) than the remaining three units. Noteworthy, one of the three axes between adjacent TMS-alkynes appears significantly more dark than the other two axes, implying that the LUMO appearance of surface-adsorbed HEB is not threefold symmetric. Since this symmetry break is consistent with our previous conclusion that assembled HEB units are chiral (enantiomorphic arrangement), we speculate that the *dI/dV* appearances of HEB within α and β domains are mirror-symmetric with respect to this axis.

5.3 Co Exposure

To study the interaction of Co atoms with the organic template in the absence of molecular diffusion, cluster formation, or catalytic reactions, we performed an in situ Co dosage with the sample kept in the STM at temperatures below 12 K. Figure 5.3a depicts a large-scale STM image with a Co coverage of approximately 0.002 monolayer (ML), whereby one ML corresponds to one Co atom per surface atom. The individual Co atoms appearing as separated sphere-shaped protrusions on the silver surface imply diffusionless Co adsorption. Comparing the density of single Co atoms on pristine Ag(111) terraces with the density of decorated molecules within the islands, we conclude that each decorated molecule incorporates one Co atom. As shown in Fig. 5.3b presenting a β domain decorated with Co, the pristine domain structure described above remains preserved due to the immobility of the molecules at 12 K, however a new species is present within the molecular matrix. This species denoted C_1 appears in six different orientations (dashed outline) and is imaged with a characteristic depression on one side and two protruding lobes at the opposite side. At such low dosage temperatures and Co concentrations, the organic template suppresses Co clustering, while at the same time intramolecular relaxation mediates the formation of the Co-bis-alkyne geometry of C_1.

Figure 5.3d shows high-resolution images of the six different orientations of C_1, which are classified by color-coded lines passing through the six possible intramolecular binding sites for one Co atom (yellow marker) within C_1. Remarkably, gas-phase DFT modeling reveals selective interactions between one Co atom and one HEB molecule, hence rationalizing the tweezer-shaped geometry of C_1. Albeit one has to be careful when comparing gas-phase configurations with STM results, the model helps in explaining its low-symmetry structure (see Fig. 5.3c for side view). Interestingly, the Co atom does not decorate the center of the substituted benzene ring, but prefers interaction with the π system of two adjacent alkynes. Such bis-alkynyl π tweezer bonding (see Fig. 1.3c) is known for various transition metals [14] and can afford stable interactions [23]. However, reports on related cobalt complexes are rare [24, 25]. Without STM image simulations, one might expect the Co atom to sit at the brighter side of the complex. However, in accordance with observations for higher Co coverage (see below), we suggest that the Co atom causes the depression and attribute the counter-intuitive STM contrast to a modified molecular conformation, where alkynes next to Co spread apart, implying that non-interacting terminal groups are pushed upward (brighter appearance).

A closer look at the two protruding lobes of the C_1 units in Fig. 5.3d reveals that one lobe appears brighter than the other, thus the complexes exhibit chirality. We define a C_1 unit to represent the 'R' enantiomer for the case that the brighter protrusion is on the right side of the main axis when going from the cobalt atom toward the brighter side. Interestingly, when analyzing the handedness of the C_1 units within the β domain depicted in Fig. 5.3b, it turns out that all complexes exhibit the same chiral nature. Consistently, within an α domain only the opposite handedness was observed (not shown). Considering the preferential adsorption of Co atoms at deeper

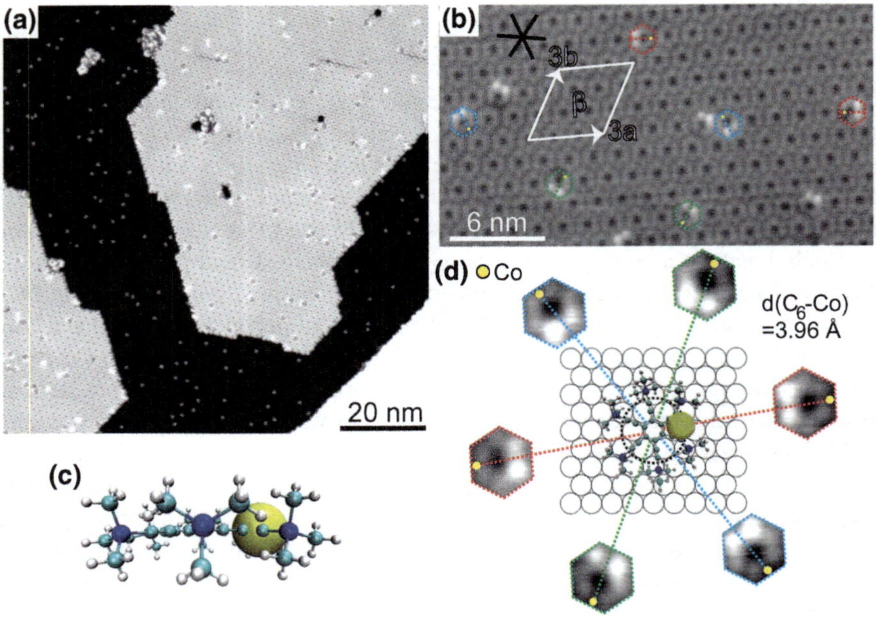

Fig. 5.3 Co-decoration of organic phase after in situ Co dosage at 12 K. **a** Large-scale STM topograph showing single Co atoms along with sub-ML adlayer of HEB. **b** β domain with Co-HEB compounds (yellow dot indicates Co atom) color-coded according to their orientations. **c** Side view of a gas-phase DFT model visualization for C_1. **d** Top view of C_1 model centered on Ag(111) registry, along with zooms of the six orientations shown in (**b**). Intersection points of dashed circle and lines coincide with six hollow sites, tentatively attributed to intramolecular Co confinement. Tunneling parameters V_t, I_t: **a** −100 mV, 300 pA; **b**, **d** −100 mV, 100 pA

lying hollow sites [26], in combination with a comparatively strong Co-bis-alkyne π tweezer binding motif (six intramolecular binding sites for Co), we centered the DFT-simulated C_1 model on the sixfold symmetric top site and aligned it according to the β domain (Fig. 5.3d). Six hollow sites (intersection points of dashed circle with lines) are encountered very close to the location of the Co atom within each of the six Co-bis-alkyne binding motifs (STM zooms) for the β domain. The alignment of the C_1 model according to the α domain (16° twist of the molecule) expresses six other hollow sites as intersection points. This symmetry observation leads us to the tentative assumption that the pristine molecules of the organic phase adsorb at top sites (cf. Fig. 5.2b). The manifestation of chirality within these surface-confined molecules and C_1 complexes supposedly results from their site-selective interactions with the Ag(111) surface. Aiming at the creation of two Co-bis-alkyne motifs within one HEB unit (dicobalt complex), the Co dosage at 12 K was increased. However, we only observed irregular Co clustering on the immobile molecular template. Next, the influence of the deposition temperature on the formation of organocobalt complexes was studied.

Co Dosage at 200 K

To investigate the influence of thermal activation on the formation and positioning of the complexes, we increased the sample temperature for Co dosage on the organic monolayer islands to 200 K. Two samples with different stochiometric ratios (Co/HEB) were prepared. The top part of Fig. 5.4 (panels a–c) refers to low Co coverage, while the data shown in panels (d–f) were obtained for high Co coverage. Already at a low Co/molecule ratio (Fig. 5.4a) one can observe the emergence of a new phase with a similar hexagonal packing scheme, but with a different unit cell containing both complexes and pristine molecules. The two domain orientations of this multi-component phase differ from the organic phase by a 30° rotation of the embedded units (highlighted with colored hexagons), hence the unit cell is rotated by the same amount. The previously introduced α (β) notation and the color scheme marking the complex's orientation are consistently used throughout this article.

In agreement with the in situ Co dosage at 12 K (cf. Fig. 5.3b, d), the organometallic complexes within one domain appear in six orientations classified by three main axis (MA) (Fig. 5.4a, b). In addition to mononuclear C_1 (dashed outline), we recognize a further species (solid outline) with two bright lobes opposite to each other and two depressions along the main axis (Fig. 5.4b) rotated by 90° with respect to the protrusions. Consistent with C_1 showing one depression, the second species is attributed to a dinuclear complex featuring two π tweezer binding motifs along the MA. Hence, the counter-intuitive intramolecular STM contrast is tentatively ascribed to a downward pulling of alkynes next to Co atoms, while non-interacting moieties are pushed upward for steric reasons and thus appear very brightly. We denote this species C_2 and highlight all Co atoms of the recognized complexes by yellow markers for better apprehension of the STM image. Figure 5.4b shows two magnified sections with superimposed DFT molecular models (gas-phase DFT model for C_1 and on-surface DFT model for C_2), which represent two complexes with the same MA. With its Co atom having two possible adsorption sites along the MA, the C_1 complex can assume two orientations with the same MA.

We observe that all complexes displaying the same MA orientation (same color) represent a subset of hexagonal lattice points. Because of three MA, the superposition of three hexagonal lattices, shifted against each other, leads to a dense-packed patterning, whereby remaining empty spaces are located at hexagonal lattice points, which according to the high-coverage sample in Fig. 5.4d are exclusively occupied by pristine molecules. Consistently with the preceding results, the periodic nanostructure is divided into four hexagonal lattices, i.e., a hexagonal lattice of pristine molecules appearing without protruding features, and coexistent with three fault-free hexagonal lattices embedding mainly C_2 complexes (solid) and a few C_1 (dashed). As shown in the close-up STM image of Fig. 5.4e where positions of Co atoms are given to guide the eye, one sees that six C_2 complexes surround one HEB molecule. The windmill-like motif is rationalized by three pairs of opposite complexes, where each pair reveals a different MA. For the sample with low Co/molecule ratio (Fig. 5.4a), most Co-HEB units are represented by C_1 replacing C_2 with same MA. Moreover, HEB molecules replace the absence of Co-HEB within the three hexagonal lattices

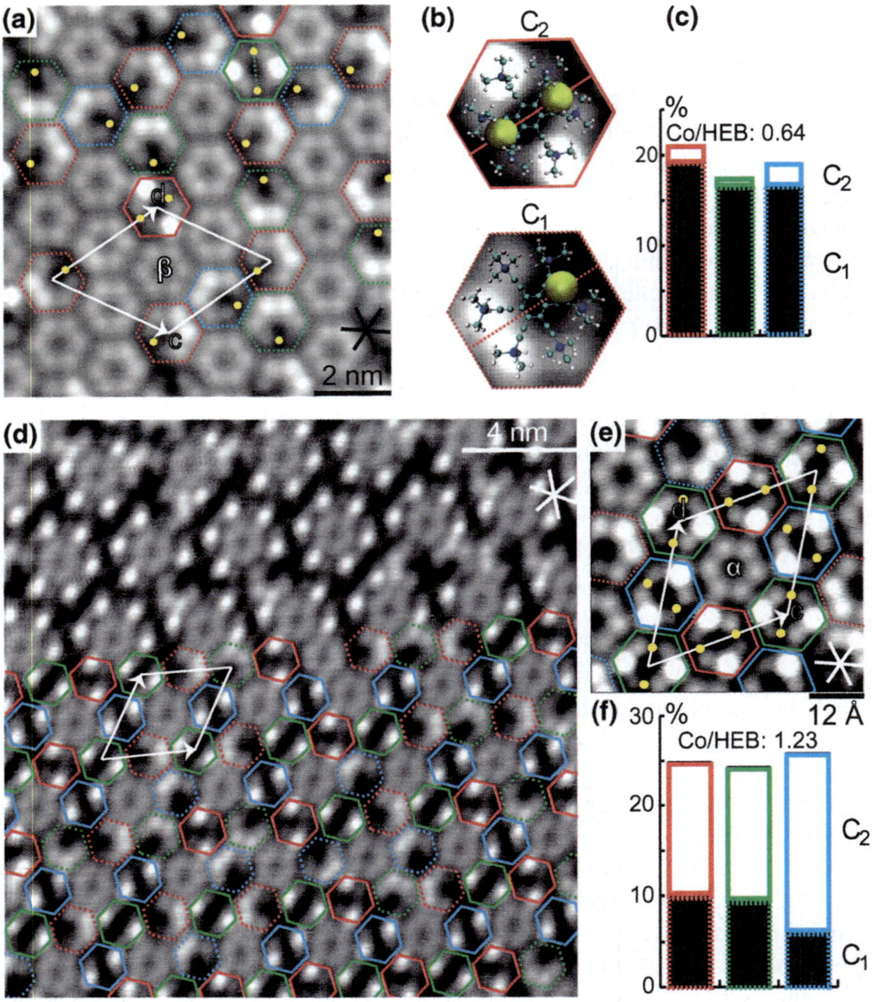

Fig. 5.4 Organocobalt phase generated via Co dosage at 200 K reveals α domain (d,e) and β domain (**a**). Close-packed directions of Ag(111) are depicted by stars. **a** Organocobalt island with a Co/HEB ratio of 0.64 together with superimposed unit cell, cobalt sites (yellow) and colored outlines marking MA. Complexes with same MA represent a subset of hexagonal lattice points. **b** Magnified STM image of two complex species with same MA within the network: mononuclear complex C_1 (dashed outline) and dinuclear complex C_2 (solid outline) incorporating one and two Co atoms along their MA (drawn line). **c** Relative occurrence of C_1 (dashed) and C_2 (solid) for each MA (ensemble of 267 units) for low-coverage case. **d** STM image of organocobalt network with a Co/molecule ratio of 1.23. **e** Pristine molecule surrounded by six C_2. **f** Relative occurrence of C_2 and C_1 (171 molecules evaluated) for high-coverage case. Tunneling parameters V_t, I_t: **a**, **b** 100 mV, 300 pA; **d** 100 mV, 100 pA; **e** 20 mV, 100 pA

of complexes. Both samples with low and high Co concentration show the same unit cell orientation and do not contain structural errors due to the misalignment or defective positioning of complexes. This implies that the strict ordering pattern is mainly ascribed to Co-HEB, while pristine HEB remains unbiased toward the partial occupation of the other three subsets of hexagonal lattice spots.

Depending on the Co concentration, a rhomboid-shaped region of four adjacent molecules within an organocobalt domain (molecular unit cell) can contain $0 \leq n \leq 6$ Co atoms; six Co atoms for the ideal case that all complexes are C_2, while other regions contain less Co atoms due to the replacement with C_1 and HEB. The relative occurrence of C_1 and C_2 within the molecular islands with low (high) Co coverage is displayed in the respective histogram in Fig. 5.4c, f, representing an evaluation of 267 (173) molecules. The similar height of the three color-coded columns indicates that all three MA are equally probable for C_1 (black fill with dashed outline) and C_2 (white fill with solid outline). In case of low Co dosage, 40% of the molecules remain pristine, while the complexes primarily represent the C_1 species (dashed outline). At high Co dosage, however, only 26% are not complexed, while C_2 is the more abundant complex species (solid outline). The strong decrease of the C_1 species further hints to the incorporation of two Co atoms within the C_2 species.

When increasing the Co/HEB ratio beyond 1.5, i.e., the value of the ideal architecture, we observe the complete dissolution of the mixed domains leaving behind isolated, intact C_2 complexes (not shown). We conclude that to induce and stabilize the formation of stable 2D organocobalt domains, at least one fourth of all constituents must remain pristine HEB (cf. hexagonal lattice of HEB monomers in Fig. 5.4d).

While deposition at 12 K leads to statistically distributed C_1 orientations due to Co adsorption on an immobile molecular template, C_1 units formed at 200 K show a strong correlation between MA orientation and adsorption site due to the mobility of HEB and Co-HEB. Optimizing the balance between intermolecular and interfacial interactions by the choice of favorable adsorption sites and alignment configurations, both C_1 and C_2 obey the same organization principle. Considering the preferential adsorption of single Co adatoms at hollow sites [26] together with the complexes' size and symmetry, and selective attachment of Co atoms via alkyne–Co–alkyne bridges, the structure of the organocobalt domains is rationalized by the site-selective positioning of Co-HEB on Ag(111). Since the pattern is independent on the distances between the complexes, which is influenced by the Co concentration (see Fig. 5.4a), we exclude that the complexes' positions and orientations are caused by electrostatic interactions between them.

5.4 Surface Tessellation

For a better understanding and atom-precise characterization of the intricate layer organization, we present high-resolution STM images with enhanced color contrast for both isolated C_2 units (Fig. 5.5a) and extended domains (Fig. 5.5b). Isolated C_2 complexes coexist as a naturally formed species with 2D networks. Due to the enhanced STM contrast (Fig. 5.5a, left panel), one recognizes an uneven apparent height of the two lobes near to the Co sites (yellow markers). The two less protruding features are connected by a white line for simple identification. This α configuration is compared to a β C_2 unit (right panel), which reflects the same MA (purple line), however, it appears slightly rotated and the intramolecular contrast of the alkynes next to Co appears inverted (mirrored white line). From the different angles between the MA and the depicted lines, we identify the chiral character of the dinuclear organocobalt complexes, labeled α and β according to their mirror-symmetry with respect to the MA. The chiral signature could be interpreted as slightly different z-heights within each alkyne pair enclosing one Co atom, and it seems remarkable that both bis-alkyne Co motifs of one complex display the same handedness. The presence of enantiomeric forms of the C_2 complex demonstrates that chirality is present at the single-complex level and is not induced by the aggregation into a 2D domain only.

To establish a direct atomic-level comparison with the two-dimensional organocobalt phase, two mirror-symmetric STM images with similar enhanced contrast are presented in Fig. 5.5b together with the underlying Ag(111) registry. Firstly, a different apparent height of alkynes interacting with Co is identified for both C_1 and C_2, and secondly, all bis-alkyne Co motifs reflect the same chirality. Therefore, we conclude that the left (right) topograph shows an enantiopure domain in which the C_2 complexes correspond to the α (β) conformer of Fig. 5.5a. The handedness of the homochiral units is invariant to rotations, thus only a mirror operation along the respective MA, which coincides with the underlying $[11\bar{2}]$ direction, causes the congruency of complexes with opposite handedness but same MA. Both isolated and laterally embedded α (β) conformers exhibit the same molecular alignment (approximately 16° twist between α and β). Compared to a racemic arrangement of α and β complexes, the organization within domains of homochiral complexes (enantiomorphic arrangement) is assumed to promote a closer domain packing, which implies stronger van-der-Waals interactions and hence a gain in adsorption and free energy.

A closer look at the proposed model for the α domain (Fig. 5.5b) reveals that molecules and complexes are aligned in such a highly ordered fashion, that all Co positions fall onto substrate hollow sites. The missing Co atoms with respect to the ideal structure (cf. Fig. 5.4e) are marked as black circles. The two Co atoms belonging to one C_2 complex always occupy one fcc and one hcp hollow site. They are connected by purple lines in order to highlight the complex's MA. Interestingly, the Co atom of C_1 is also located at either fcc or hcp hollow site, indicating that the presence of C_1 does not disturb the highly organized superlattice of incorporated Co atoms.

When linking the fcc (hcp) hollow sites of three adjacent complexes by an fcc (hcp) triangle and further connecting an fcc and hcp triangle by a rhomboid, a 3.4.6.4 tiling pattern [27] is clearly evident (lower section of Fig. 5.5b). As highlighted by the colored polygons, one large hexagonal tile is surrounded by six triangles, three oriented upward (green) and three oriented downward (brown), and six rhomboids (orange). The unit cell contains one hexagon, two antiparallel triangles and three 120°-multiple rhomboids. The corners of the polygons are situated at hollow sites

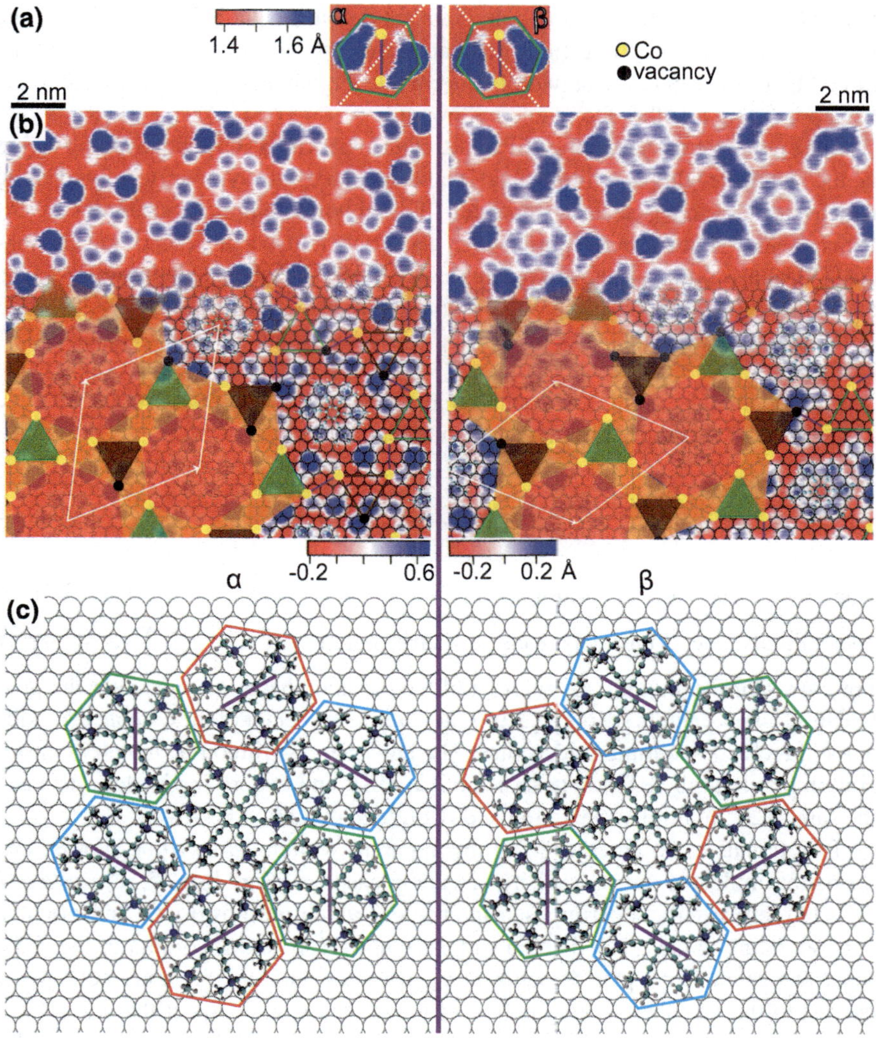

Fig. 5.5 STM images with enhanced color contrast display isolated C_2 conformers and organocobalt domains. **a** Two isolated C_2 complexes exhibiting mirror symmetry with respect to their MA (purple line) represent the α and β conformer, respectively. **b** Atom-precise scrutinization of enantiopure α and β domain with underlying Ag(111) lattice and superimposed surface tesselation scheme. Co sites (yellow) and vacancies (black) of C_1 and C_2 coincide with fcc and hcp hollow sites. Threesomes of complexes are grouped by fcc and hcp triangles (color-coded), which together with rhomboids (orange) and hexagons (red) reflect a 3.4.6.4 surface tessellation pattern. **c** Adsorption model for windmill motif (cf. Fig. 5.4e) based on Ag(111) registry and molecules arranged according to (**b**). Straight lines connecting opposite hollow sites (purple) coincide with MA orientations in (**b**) (purple-colored Co–Co axis). Tunneling parameters V_t, I_t: **a** -10 mV, 30 pA; **b** left: -10 mV, 100 pA; right: 20 mV, 100 pA

where Co is located. Comparing both domains (left and right STM topograph), one notes that the superimposed tiling scheme is mirror-symmetric with respect to the three MA (purple lines). The twofold degenerate alignment of C_1 is attributed to the Co atom sitting at either fcc or hcp hollow site.

The mathematical description of this structure dates back to Johannes Kepler, who deduced that the complete filling of an Euclidean plane with symmetric polygons can lead to just eleven tessellation structures [28]. Hitherto, numerous examples of surface tessellation have been reported, e.g., a Kagomé lattice using dicarbonitrile linkers [29], snub-square tiling within a lanthanide-based metal-organic framework [30] and a rhombitrihexagonal tiling in a solution-confined oligophenyleneethylene-based Pd(II) complex [31]; the latter motif is equivalent with our surface-confined tessellation pattern.

From a detailed analysis of both domains, we deduce a commensurate $\sqrt{129} \times \sqrt{129}$ R 22.4° superstructure with the α unit cell reading,

$$\begin{pmatrix} \vec{a} \\ \vec{b} \end{pmatrix} = \begin{pmatrix} 8 & 5 \\ -5 & 13 \end{pmatrix} \begin{pmatrix} \vec{u} \\ \vec{v} \end{pmatrix},$$

whereby $\vec{u}$ and $\vec{v}$ are the primitive vectors of Ag(111) depicted in Fig. 5.2b. Correspondingly, the β unit cell reads in matrix notation: $\begin{pmatrix} 13 & -5 \\ 5 & 8 \end{pmatrix}$, as shown in Fig. 5.4a. Consistent with our analysis of the organic phase, we encounter the pristine molecules' centers adsorbed at top sites, while the center of the complexes reside at the three 120°-multiple bridge sites. The latter agrees with our symmetry consideration that both the STM appearance of Co-HEB and underlying bridge site share the same 120°-multiple orientation and twofold symmetric axis (MA).

To rationalize the emergence of the apparent windmill-like tiling motifs, i.e., a pristine molecule surrounded by six complexes (cf. Figs. 5.4e, 5.5c displays a Ag(111) lattice with superimposed molecules, which are positioned and oriented corresponding to the α (β) domain's registry in the left (right) column of Fig. 5.5b. Since we encounter a pair of opposite hollow sites, i.e., one fcc and one hcp site nearest to the gas-phase configuration of the complex, we suspect that the MA of an organometallic unit is determined by the respective alignment of the underlying bridge site. Assuming that the orientation of the complex is correlated with its adsorption site, the intriguing surface tessellation pattern can be explained. In the case of a dense-packed molecular matrix and sufficient Co atoms, three fourths of the molecules can form twofold symmetric complexes, which will be centered at bridge sites. The remaining one fourth of molecules is necessary to complete the remaining voids with the sixfold symmetric precursors' centers located on top sites, which would be unfavorable for the complexes.

5.5 Modeling of Organocobalt Phase

For a characterization of the extended supramolecular tessellation from a theoretical point of view, state-of-the-art DFT calculations including the Ag(111) substrate were performed by Dr. Ari P. Seitsonen [32]. The input of this DFT calculation is based on positioning Co atoms on a molecular lattice according to the experimentally observed organocobalt phase. To check reproducibility, the calculations were carried out with two different approximations of the exchange and correlation term. Both the revPBE+D3 and B86r-vdW-DF2 treatment provide very similar results, therefore only the results with the former procedure are presented (for details on the calculations, please refer to the published manuscript).

A visualization of the lowest energy geometry is presented in Fig. 5.6. To facilitate comparison with the experimentally observed structure (cf. Fig. 5.5), the Ag atoms in Fig. 5.6c are represented by red (instead of grey) spheres. Notably, the DFT results indicate the on-surface formation of a dinuclear compound with a Co–Co distance of 7.76 Å (Fig. 5.6b). Since this distance is close to twice the Co-arene distance (3.96 Å) of the gas-phase DFT model of C_1 (cf. Fig. 5.3d), we postulate a minor influence of the Ag(111) surface on the Co-arene distance during $C_1 \rightarrow C_2$ conversion.

The DFT structure of the organocobalt domain has a unit cell with six Co atoms (yellow) and four molecules (white outline in Fig. 5.6c), consistent with the observed α domain (cf. Fig. 5.4e and left part of Fig. 5.5b). In agreement with experiment, molecules without Co atoms are centered on top adsorption sites, while complexes are centered at bridge sites, thus establishing the observed windmill-like motifs. With the Co atoms preferentially binding close to fcc (hcp) hollow sites and representing the corners of superimposed brown (green) triangles, the experimentally-observed tessellation pattern is reproduced.

Comparing the DFT results with the STM contrast of the complexes, the height of the topmost Ag atoms of the five-layer Ag(111) slab below the molecules is analyzed in Fig. 5.6a, where color-coding and scaling refers to labeled z-height intervals with z = 0 representing the equilibrium height of the topmost layer. The Ag atoms with the lowest (highest) z values are exclusively encountered below the complexes, i.e., two red (blue) Ag atoms along the Co–Co axis (perpendicular to it). Specifically, the Ag atom below each alkyne–Co–alkyne motif is indented (red), while the Ag atom below each non-interacting TMS–alkyne moiety of C_2 is protruding (blue). Since the surface corrugation of the topmost Ag(111) layer is absent below the HEB molecules, we assume that it originates from the embedded Co atoms of the complexes. We postulate that they are pulled down via Co atoms, by which a strong interaction between alkynes and Ag atoms is only present for Co-HEB. For each complex, the d orbitals of each of the two protruding Ag atoms (blue) might strongly hybridize with the out-of-plane π orbitals of the overlying alkyne. The remaining four alkynes' in-plane π orbitals are supposed to strongly hybridize with the d orbitals of Co instead, wherefore the Ag atom below each alkyne–Co–alkyne motif (red) is lowered. The periodicity and distribution of the surface corrugation (0.58 Å height difference between blue and red spheres) agrees qualitatively with the

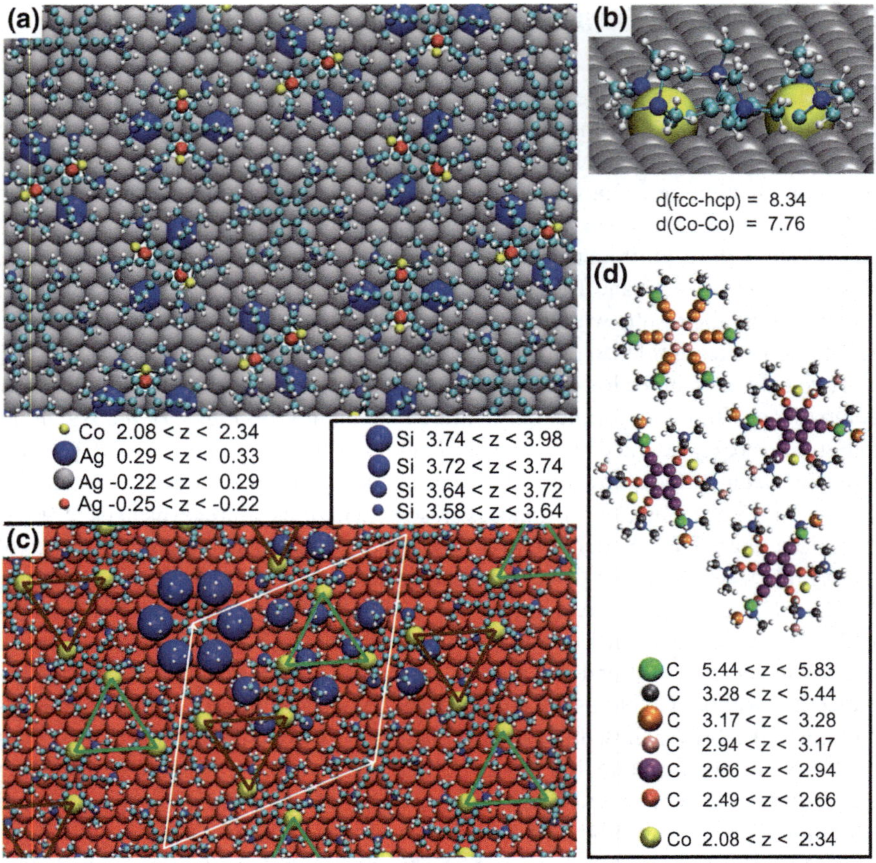

Fig. 5.6 DFT-modeling on organocobalt α domain. **a** DFT model emphasizing the corrugation of the topmost Ag(111) layer, as indicated by z-height intervals and scaled spheres for indented (red) and protruding (blue) Ag atoms. **b** Side view of single C_2 with Co–Co and fcc–hcp hollow distance. **c** The relaxed geometry displays a superlattice of embedded Co atoms, which were grouped by fcc and hcp triangles. The varying z-height of the Si atoms is illustrated by scaled spheres (blue) within the unit cell. **d** Physical heights of all carbon atoms within molecular unit cell (color-coding according to labeled z-height intervals). Diameter of Co atoms is reduced in (**a**) and (**d**) to enhance visibility of Ag and C atoms, respectively. All numeric values are in Å

STM contrast observed in Fig. 5.4e, where the complexes' non-interacting moieties show the strongest brightness, while the darkest contrast is found at the Co sites. From DFT-calculations, similar structural arrangements of the topmost surface layer were reported for a charge-transfer system with electron donation from Cu(100) to an organic layer of assembled tetracyanoquinodimethane molecules [33].

Based on the surface corrugation implying an increased surface-molecule interaction for Co-HEB, we expect a lower z-height compared to HEB. Indeed, when comparing the physical heights of all carbon atoms (Fig. 5.6d), the interacting C(sp)

atoms of Co-HEB are lowered by 0.65 Å compared to the C(sp) atoms of HEB. As a consequence of this alkyne–Co–alkyne anchoring at hollow sites, the arene ring (C(sp2) atoms) of Co-HEB is lowered by 0.26 Å. These molecular conformations brought about by the anchoring via Co atoms are seen as strong bending of the alkyne-TMS moieties next to Co (side view in Fig. 5.6b).

Comparing the heights of the Si atoms, as illustrated by the scaled diameters of prominent blue spheres within the unit cell in Fig. 5.6c, one notices the equal z-height of the six Si atoms of pristine HEB. On the contrary, for C_2, the two Si atoms of the TMS–alkyne–Co–alkyne–TMS binding motifs systematically display different heights, which implies a non-uniform Co-bis-alkyne interaction causing a significant bending of the respective alkyne units. Consequently, the DFT calculation confirms the formation of an enantiopure organocobalt domain and rationalizes the chiral character of C_2 by the bending of the two TMS–alkyne groups next to Co. Although the physical heights of the simulation image cannot be directly related to the STM contrast of Fig. 5.5b, it should be mentioned that the TMS–alkyne being closer to Co appears brighter in the STM contrast, while the DFT-image represents it with a smaller Si height than the less interacting moiety of the alkyne–Co–alkyne motif. The fact that the TMS groups of HEB appear less bright (within the STM contrast) than the non-interacting TMS groups of Co-HEB is not reflected within the DFT visualization of the Si and C heights (Fig. 5.6c, d), but explained by the structural arrangement of the topmost surface layer, i.e., lifting and lowering of topmost Ag atoms below the complexes (Fig. 5.6a). Since the STM contrast reflects a convolution of both the z-height and electronic DOS, discrepancies related to the Si/C heights can be related to hybridization effects with the substrate.

A difference between the DFT structure and the model derived above from our STM observations is that the molecules are slightly rotated with respect to the experimental α domain configuration. This might be due to a limited accuracy of the DFT results in reproducing the delicate balance between Ag-alkyne and alkyne–Co–alkyne interactions.

5.6 Single-Molecule Manipulations on Organocobalt Complex

To inspect the structural integrity and conformational flexibility of isolated complexes, we performed single-molecule manipulation experiments. The STM image in Fig. 5.7a shows small assemblies together with naturally formed isolated C_2 units. The histogram in the inset displays the distribution of their MA orientations within a region of $200 \times 200\,\text{nm}^2$. It reveals that all three MA coexist with nearly equal frequency within the statistical error. Figure 5.7b depicts STM images before and after a 120° rotation induced by the STM tip. For this purpose, the tip was centered on C_2, approached in the OFL mode (initial setpoint: $V_t = -10\,\text{mV}$, $I = 50\,\text{pA}$) until current spikes indicate the rotation of C_2, in response to the increased tunneling current

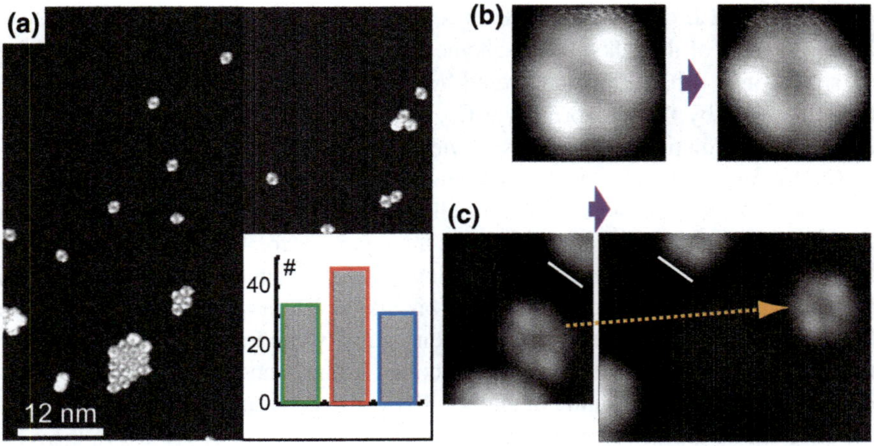

Fig. 5.7 STM-tip manipulations on isolated C_2 complexes. **a** STM image illustrating isolated C_2 units distributed over a Ag(111) terrace. Inset: Histogram of MA orientations. **b** STM images after tip manipulation showing a 120° rotated species. **c** Sequence showing a tip-assisted displacement of C_2 (along the depicted line) and associated rotation by 120°. Tunneling parameters for imaging V_t, I_t: **a–c** −10 mV, 50 pA

(typically at $I = 20\,\text{nA}$). The similarity of the molecular features before and after the rotation hints to the structural integrity of C_2 upon this kind of tip manipulation.

The complexes' stability toward high tunneling currents, as long as the bias voltage remains low, was further used to demonstrate their cohesion and intactness toward lateral displacements. Figure 5.7c presents STM images of the complex before and after the lateral movement along the marked line. This displacement was achieved by approaching the tip until reaching $I_t = 25\,\text{nA}$ and dragging the complex along the path of the tip (orange arrow) with respect to a fixed reference mark (white line). While attemps in manipulating pristine molecules resulted in molecular adsorption on the tip, isolated complexes could easily be moved without triggering any structural modification, as evident from the consecutive STM images. The latter implies a stronger interfacial coupling for Co-HEB compared to HEB. The 120° rotation of the complex after the manipulation hints to a correlation between its MA and adsorption site (cf. Fig. 5.5c). Therefore, if the next-nearest bridge site at the final position is not parallel to the complex's MA, it is supposed to rotate by 120° during migration to this bridge site with favorable $[11\bar{2}]$ orientation, thus allowing the Co atoms to reach the preferred hollow sites. Only in the case that the orange arrow is parallel to the $[11\bar{2}]$ direction of C_2's initial adsorption site in the left STM image of Fig. 5.7c, the migration should proceed without any 120° rotation.

Kondo Signature of Activated Complex

Complexes incorporating multiple magnetic metal centers [3, 34, 35] are discussed as promising candidates for nanomagnets, spintronic and quantum computing devices [36–38]. The research focus lies particularly on surface-supported molecular magnets expressing long magnetic lifetimes [39–41] and remarkable hysteresis characteristics [42]. Accordingly, we explored the properties of isolated complexes regarding potential magnetic behaviour. Initially, we performed STS measurements on isolated C_2 units, which do not show a Kondo feature. Then, we applied a stronger manipulation procedure than before, in order to alter the interaction between Co atoms and alkynes: the STM images I and II of Fig. 5.8a were obtained after applying a bias pulse of −2 V for a duration of 10 ms on top of the left bright lobe (top image). The structurally modified complex I is characterized by a slight rotation, one remaining protrusion and a grainy frizzled feature (red outline) located between two alkynes. The manipulation experiments were repeated for different complexes and show that configuration I is highly reproducible. Occasionally, a voltage pulse on a protrusion triggered the emergence of a strongly rotated compound with two grainy frizzled features, as identified by two outlined spots in configuration II. A further voltage pulse on configuration II converted it back to I, identifying the latter as more stable.

Interestingly, the frizzled features of the I and II variants in Fig. 5.8a appear at different locations compared to the Co sites of the initial structure above. We attribute them to loosely bound Co atoms losing their initial binding sites and changing to a metastable binding configuration, in which they partially follow the STM tip. In configuration I, one of the two Co atoms has undergone this site change, while

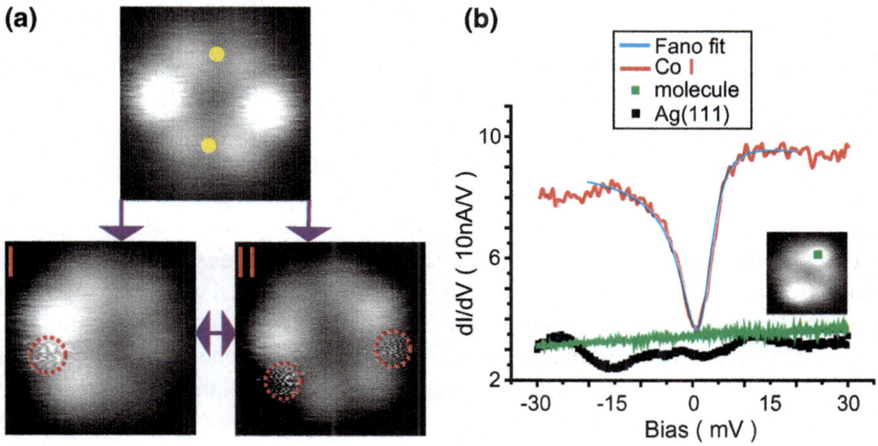

Fig. 5.8 Kondo features in activated isolated C_2 species. **a** Activation of pristine complex (top image) by a bias pulse (−2 V, 10 ms) with the tip positioned above a bright lobe. Occurrence of one (two) frizzled spots (red outline) in activated species I and II is accompanied by the reduced brightness of one (two) protrusions and a rotation of HEB. **b** Low-bias *dI/dV* spectra on pristine Ag(111) (black) and C_2 [green: spectrum on pristine complex acquired at position of green marker in the inset in (**b**), and red: spectrum acquired within red circle of configuration I in (**a**)]. The Fano fit (blue) was carried out with $q = 0.27$, $x_0 = 1.9$ mV, $T_K = 23$ K. Tunneling parameters for imaging V_t, I_t: **a**, **b** −20 mV, 100 pA. STS setpoint parameters: −20 mV, 100 pA (120 pA for Kondo spectrum), $\Delta z = -90$ pm before sweep, $V_{mod} = 1$ mV rms, $f = 933.5$ Hz, $T = 1.2$ K

in configuration II, both Co atoms have changed their adsorption sites. While the transformation from I to II is reversible, we could not convert the activated complex to pristine C_2.

When performing *dI/dV* spectroscopy above a frizzled feature of the activated configuration I (red outline, Fig. 5.8a), we observed a strong Kondo resonance. The low-bias STS spectrum exhibiting a dip-like structure is shown in Fig. 5.8b together with *dI/dV* spectra on pristine Ag(111) and C_2, both revealing a featureless shape (STS data obtained at 1.2 K). We suggest that the dip feature indicates the existence of a magnetic moment that is screened by the surrounding electrons giving rise to a many body state [43]. The shape of the Kondo resonance shown in Fig. 5.8b has been reproduced on many activated C_2 species (configuration I) with different STM tips. It was fitted by the Fano equation

$$dI/dV \propto \frac{(q\Gamma + E - E_0)^2}{\Gamma^2 + (E - E_0)^2},$$

whereby E_0 and Γ are the position and temperature-dependent half width at half maximum of the Kondo feature and q is the shape parameter giving different weight to the interfering quantum-tunneling paths. After eliminating thermal broadening and modulation voltage (root mean square values V_{rms}) broadening by the formula

$$\Gamma(T) = \sqrt{(3.5k_B T)^2 + (\sqrt{6}eV_{rms})^2 + 2(k_B T_K)^2} \tag{5.1}$$

with k_B and e being the Boltzmann constant and the elementary charge, respectively [44, 45], the obtained Kondo temperature reads T_K =23 K. This value is remarkably lower than for individual Co atoms on the Ag(111) surface (95 K) [46]. The diminished value could be related to Co-alkyne orbital interactions weakening the Co-surface bond and thus the Kondo screening, which originates from the delocalized cloud of surface electrons. Configuration II (two grainy spots), which was rarely observed, also expressed a Kondo signature on both sides, but with different Kondo temperatures. Similar activation of C_2 entities embedded within organocobalt islands also triggers the emergence of a frizzled feature displaying a Kondo resonance. This signature on tip-manipulated complexes reveals a chemical fingerprint of incorporated Co atoms, which further consolidates our interpretation of an on-surface formed dinuclear organocobalt complex.

Conformer Switching

Single-molecule switches and molecular motors, which are powered by the tunneling electrons of an STM tip [47–50] or its electrical field [51, 52] bear high potential for nanoelectronic devices. Furthermore, STM represents an exquisite tool to identify switching between enantiomeric states of surface-adsorbed chiral molecules, e.g., the vibration-induced switching of propene [53].

Intrigued by the existence of the different conformer states (α and β) of the C_2 complexes, we explored the possibility to switch between C_2 enantiomers with the

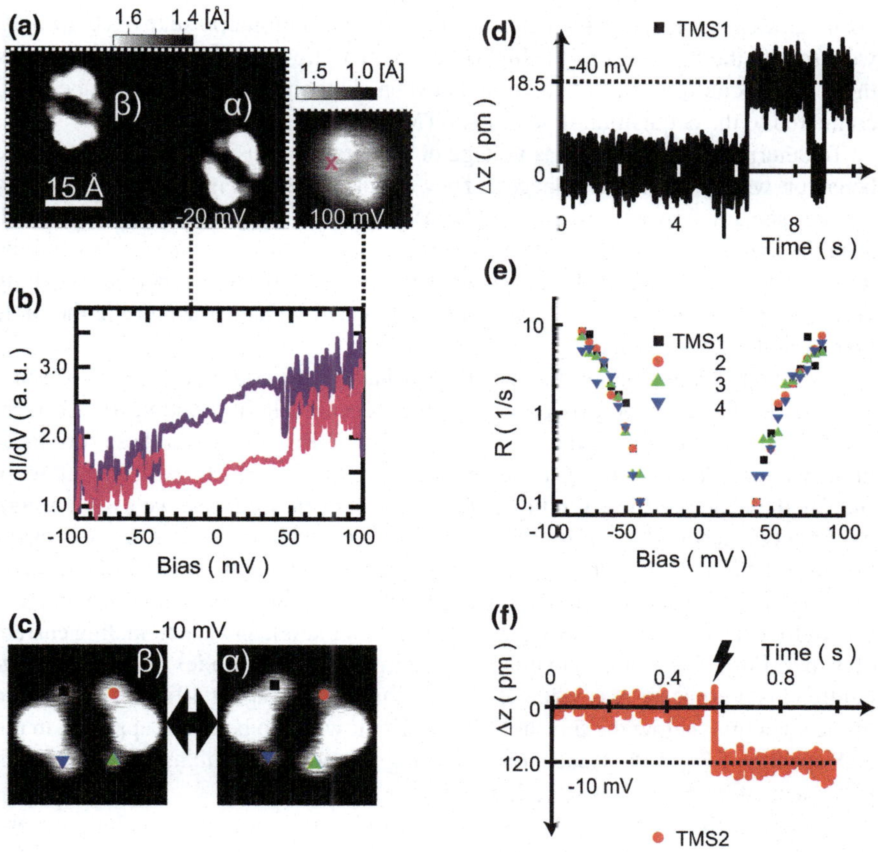

Fig. 5.9 STM and STS results qualifying isolated C_2 as a chiral complex. **a** STM topograph at −20 mV allows differentiation of the α and β conformer, while the adjacent STM image at 100 mV appears frizzled. **b** Two subsequent *dI/dV* spectra obtained at spot marked in (**a**) exhibit two symmetric plateaus below a threshold of 40 mV and two-state fluctuations above it. **c** Enhanced-contrast STM images before and after switching the chiral state, whereby marked molecular lobes flipped their brightness contrast. **d** Δz(t) spectrum on top of marked lobe for a bias voltage above threshold. Abrupt changes between plateaus indicate switching events. **e** Logarithmic plot of switching rate as a function of the bias voltage for TMS 1, 2, 3, 4 (markers). **f** Controlled switching by applying a short bias pulse (−75 mV) while monitoring z-height below threshold voltage. Tunneling parameters for imaging V_t, I_t: **a** −20 mV, 100 pA; inset: 100 mV, 300 pA; **c** −10 mV, 300 pA. STS setpoint parameters: **b** 100 mV, 300 pA, $V_{mod} = 1$ mV rms, $f = 733.5$ Hz, $T = 4.5$ K. Time sequences at $T = 1.2$ K: **d** −40 mV, 30 pA; **e** 30 pA; (f) −10 mV, 50 pA

aid of the STM tip. The STM image in Fig. 5.9a displays two isolated C_2 units, which, according to the brightness of the four alkynes next to Co, are recognized as β and α conformers (cf. Fig. 5.5a). When imaging the α complex with an elevated tunneling bias ($V_t = 100$ mV), we observe a fuzzy appearance of isolated C_2 (right panel of Fig. 5.9a). In subsequent imaging with low bias voltage (∼20 mV) the same complex again appears stable and with a clearly discernible enantiomeric state. Accordingly, we attribute the fuzziness of the high-bias image to tip-induced switching between the α and β conformers, i.e., the time between switching events seems to be small compared to the acquisition time of the STM image.

To determine the critical bias voltage of this drastic transition in the appearance behavior, we recorded *dI/dV* spectra. They are displayed in Fig. 5.9b and show two regimes separated by a threshold of ± 40 mV. For bias voltages below the threshold, a stable signal with either low or high conductance is present, whereas beyond the threshold a fluctuating signal is observed. The latter is assigned to repeated switching between the two enantiomeric states causing low and high conductance at the same tip position (cf. Fig. 5.9c).

For better characterizing the switching behavior, we recorded a series of spectra over all TMS-alkyne groups involved in organometallic bonding (markers in Fig. 5.9c), which are referred to as TMS 1, 2, 3, 4 in Fig. 5.9c–f. At these four locations, we recorded time traces of the tip height changes with closed feedback loop, such as the example shown in Fig. 5.9d. For each of these traces, we assign abrupt jumps between two plateaus of different z-height to switching events between enantiomeric states and determine the switching rate by dividing the number of events by the acquisition time (10 s). The dependence of the switching rate on the bias voltage is shown for each of the four TMS–alkyne groups (markers), and at a tunneling current of 40 pA (Fig. 5.9e). For all moieties, the logarithmic scale implies that an exponential trend is present for bias voltages above a threshold of about 40 mV. The absence of the switching below the threshold is consistent with the behavior apparent in the STS spectra of Fig. 5.9b. Hence, the switching is associated with an electron-driven process exciting molecular vibrations.

To control the switching process without triggering multiple switching events, a short bias voltage pulse (-75 mV and 45 ms) was employed while acquiring z(t) for a setpoint voltage below the threshold. As shown in Fig. 5.9f, the apparent edge due to the pulse coincides with a jump in z, indicating a controlled switching event. Recording STM images before and after such switching demonstrates that a single switching event suffices to reproducibly alter between the two enantiomeric states of a C_2 unit. All four interacting TMS–alkyne moieties (marked molecular lobes in Fig. 5.9c) change their brightness after one switching event: diagonally opposite lobes appear with same brightness while adjacent lobes reveal an apparent height difference. Hence, we assume that the out-of-plane bending (oscillation) of these intramolecular lobes is coupled. For each two-state switching process of the isolated C_2 species, all four interacting groups change z-height in a synchronous way, whereby two diagonally opposite moieties oscillate in phase (same z-height), while two adjacent lobes express opposite conformation states (opposite z-height).

In contrast to isolated C_2, it was not possible to switch the chirality of C_2 units embedded in the 2D organocobalt islands. According to our interpretation, a switching process represents an interconversion of the complex's handedness which is accompanied by a rotation of the molecule (cf. Fig. 5.9c). Presumably, this process is prohibited due to steric interactions with surrounding molecules within the dense-packed arrangement of the enantiomorphic phase. The switching of isolated Co-HEB shows the relationship between its two chiral states and the respective unit cell orientation of assembled domains. Both the isolated compounds as well as unit cell of the organic and organocobalt phase reveal a twist of ~16° between α and β. During $\alpha \leftrightarrow \beta$ switching, the Co atoms along the MA of isolated C_2 may remain at fixed locations, or rotate (together with their molecular host) symmetrically to the complex's underlying bridge site. The latter scenario would be more consistent with the observed 120°-multiple rotations of the Co–Co axis (MA) (cf. Fig. 5.7b, c).

5.7 Organolanthanide Complexes

As exemplified by the presented Co-HEB compounds, a metal alkynyl complex with multiple binding sites can serve as a chemical pocket for the intramolecular confinement of single atoms. To demonstrate the diversity of this class of on-surface complexes, it would be interesting to study how guest species with larger ionic radii compared to the 3d transition element Co (effective ionic radius depending on charge and spin: 0.53–0.75 Å) influence the organometallic bonding characteristics. To this end, we utilize the 4f-block element holmium (Ho) expressing a nearly twice as large atomic radius (effective ionic radius depending on coordination number: 0.90–1.22 Å) compared to Co [54, 55]. The general properties on lanthanide elements will be briefly mentioned in the next chapter on a Ho-catalyzed coupling reaction. Due to its large atomic size, we assume that the Ho atom does not fit within the intramolecular region between two adjacent alkynes (cf. with Co-HEB model in Fig. 5.6b, c). Thus, we expect the formation of a different complex geometry after dosing concerted amounts of Ho on a sample with predeposited HEB.

Similar to the preparation of Co-HEB, the on-surface fabrication of Ho–HEB complexes is accomplished by creating an organic layer (sub-ML coverage at $T_{deposition} = 450\,\mathrm{K}$), which is subsequently exposed to a beam of Ho atoms at a sample temperature (200 K), at which molecules and Ho atoms are mobile. Figure 5.10a depicts a large-scale STM topograph of a molecular domain incorporating a complex species that appears substantially more bright than pristine HEB. The color-coding of the depicted outlines emphasizes the six orientations of this organoholmium complex comprised of one HEB monomer and one Ho atom (denoted C_1). When superimposing gas-phase DFT models for HEB and spheres for Ho atoms (Fig. 5.10d) to the high-resolution STM image in Fig. 5.10c, one sees that the four bright lobes of C_1 are localized at TMS–alkyne groups. The spherically-shaped protrusion inbetween two adjacent alkynes within C_1 is attributed to one Ho atom. Consistent with the organometallic alkyne–Co–alkyne tweezer motifs within Co-HEB, the Ho atom's off-centered location implies that π-like metal-alkynyl interactions dominate over bonding with the six $C(sp^2)$ atoms of the central arene ring of HEB. Since the large diameter of Ho inhibits its incorporation between two alkyne groups, it is supposed to reside either below or above the molecular plane (cf. Fig. 1.3c). Remarkably, the unit cell of the organoholmium phase (α domain in Fig. 5.10c) is congruent with the unit cell of the organic phase (cf. Fig. 5.2b). This scenario contrasts the case for

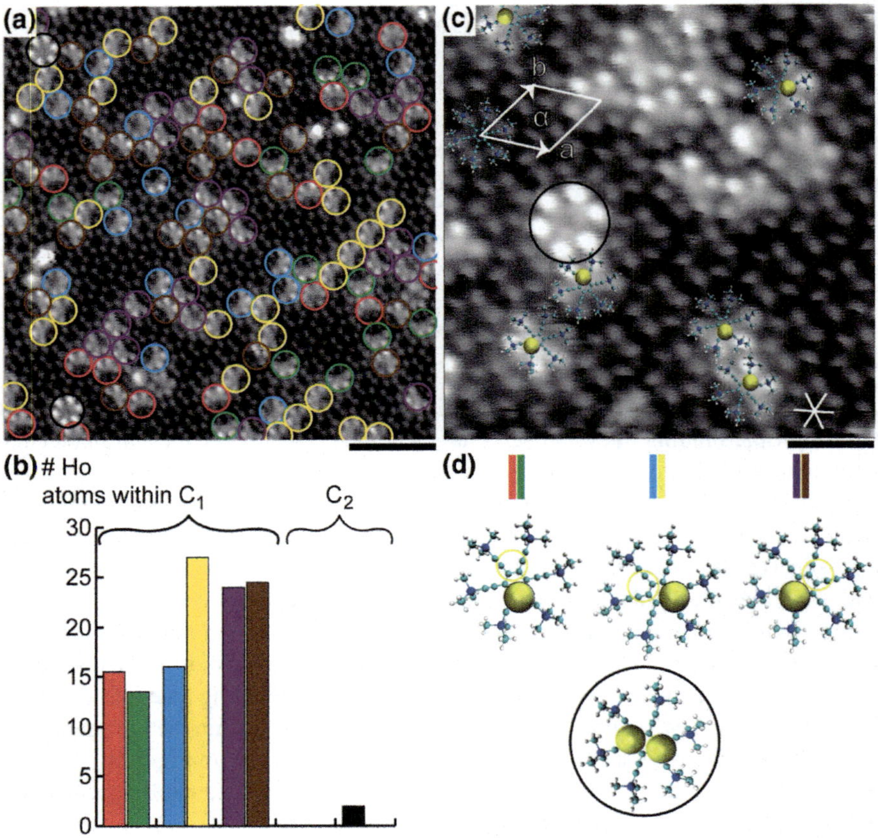

Fig. 5.10 STM results for sample preparation with low Ho/HEB ratio (Ho dosage at 200 K). **a** Large-scale STM topograph displaying assembled Ho–HEB complexes. **b** Histogram depicting statistical evaluation of their six orientations (color-coded outlines for C_1) in (**a**). **c** High-resolution STM image of domain with superimposed models for C_1 (unit cell α and high-symmetry axes (star) are indicated). **d** Gas-phase models for HEB with on-top positioned Ho atoms (yellow spheres) represent C_1 and C_2. C_2 is marked by black circle in (**a**, **c**, **d**). Tunneling parameters V_t, I_t: **a** 100 mV, 50 pA; **b** −10 mV, 50 pA. Scale bars: **a** 5 nm; **b** 2 nm

Co-HEB, in which Co deposition on a molecular adlayer steers the creation of a phase with 30°-rotated unit cell compared to the organic phase. The related 30° rotation of HEB and Co-HEB within their mixed assemblies was explained by the synergy of site-selective alkyne–Co–alkyne bonding and preferential Co adsorption near to hollow sites.

Interestingly, the mononuclear Ho–HEB units within the dense-packed molecular matrix (outlines) are arranged in a branched structure, where each compound is adjacent to one or more complexes ($1 \leq n \leq 5$ next-nearest C_1 units). This structural arrangement identified by LT-STM imaging contrasts the periodic 2D array of Co-HEB complexes within organocobalt domains (cf. Fig. 5.4d). Unlike the Archimedian tesselation structure expressed by the embedded Co atoms of Co-HEB (cf. Fig. 5.5b), the superlattice of Ho atoms within the organoholmium phase is not periodic. Accordingly, we assume that the Ho atoms within Ho–HEB are not sensitive to the atomic registry of Ag(111), suggesting that they are well decoupled from site-selective surface interactions due to adsorption on top of the HEB molecule. As opposed to the scenario for Co-HEB (Co favoring hollow adsorption sites), we hence exclude Ho confinement below the molecular plane.

However, the Ho atoms express site-selective intramolecular interactions within the sp-hybridized periphery of HEB. Similar to the C_1 species of Co-HEB, there are six orientations of Ho–HEB, which correspond to six possible binding sites for a Ho atom. As indicated by three molecular models in Fig. 5.10d (yellow circles depict Ho sites of the remaining three C_1 orientations rotated by 180°), the Ho atom is supposed to sit on top of the molecule about midway between two adjacent alkyne units. The asymmetric STM contrast due to the enhanced brightness of four TMS–alkyne groups next to the Ho-related protrusion of Ho–HEB is explained by their upward bending due to organometallic interactions between the alkynes' π orbitals and off-centered Ho atom. Thereby, the Ho atom establishes a tweezer-like bonding geometry different from Co-HEB (see Fig. 1.3c (right part)). Based on the aforementioned adsorption behavior of the organoholmium phase and comparison with the organocobalt phase, we tentatively exclude the possibility that a Ho atom located below the molecular plane causes the asymmetric upward bending of four TMS–alkyne moieties of C_1.

Besides C_1, a further Ho–HEB species reflecting a symmetric STM contrast is observed (black circles in Fig. 5.10a, c). Due to its rare occurrence at low Ho–HEB ratios, we assign it to a dinuclear Ho–HEB species and denote it C_2. While C_1 gives rise to four bright molecular lobes, the symmetric STM contrast of C_2 is interpreted as the lifting of all six TMS–alkyne groups in response to the presence of two opposite Ho atoms within the chemical pocket of HEB. The C_2 model in Fig. 5.10d (within black circle) represents only one of three 120°-multiple alignments for a pair of opposite Ho atoms, each confined in a tweezer-like fashion. For comparison with Co-HEB, the STM contrast of C_1 and C_2 is explained by the downward (upward) bending of interacting alkynes toward the lower lying Co (upper lying Ho) sites. Due to the symmetric STM appearance of C_2, the alignment of the Ho-Ho axis cannot be resolved. Besides conformational changes of HEB in response to Ho-alkyne interactions, also interfacial and intermolecular charge-transfer processes may contribute to the STM appearance of C_2.

The statistical distribution on the statistical frequency and orientation of Ho–HEB is presented by the histogram in Fig. 5.10b, whose color-coding corresponds to the C_1 orientations encountered in the STM image of Fig. 5.10a. From the 124 Ho atoms decorating molecules within a region of $25 \times 25\,\mathrm{nm}^2$, we conclude a Ho coverage of $\sim$0.014 ML within the decorated island. At this Ho concentration, only a few C_2 units are encountered, while the six orientations of C_1 are almost equally frequent within the statistical error.

To further study the details on the multiple bonding aspects of Ho–HEB, we prepared a sample with higher stochiometric Ho/HEB ratio by dosing more Ho (same deposition temperature). The high-coverage sample shown in Fig. 5.11a con-

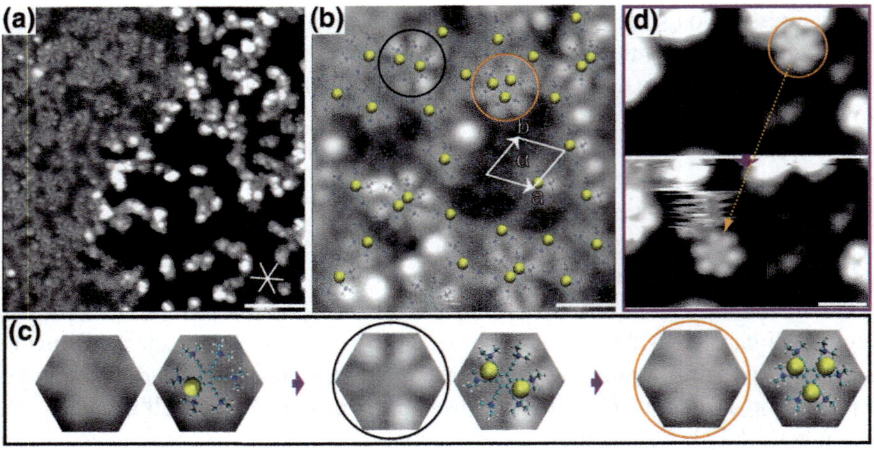

Fig. 5.11 STM results for sample preparation with high Ho/HEB ratio (Ho dosage at 200 K). **a** STM image showing organoholmium phase coexisting with isolated complexes next to Ho clusters. **b** High-resolution STM topograph with superimposed HEB models (semitransparent), Ho atoms (yellow) and depicted unit cell vectors (white). **c** Zoomed STM images and superimposed models schematically depict C_1, C_2 (black outline in (**b, c**)) and C_3 (orange outline in (**b–d**)). **d** STM images before (top) and after (bottom) lateral manipulation on isolated C_3 complex. Tunneling parameters V_t, I_t: **a–c** −100 mV, 50 pA; **d** −100 mV (top) and −10 mV (bottom), 50 pA. Scale bars: **a** 8 nm; **b, c** 2 nm

tains four times the amount of Ho compared to the low-coverage sample discussed above, wherefore a significantly higher C_2/C_1 ratio is present. When superimposing a semitransparent model for the pristine molecular layer comprised of HEB units (Fig. 5.11b), one clearly recognizes the embedded Ho–HEB units within the assembled island expressing the same unit cell orientation as the pristine phase. According to the asymmetric and rather symmetric STM appearance of C_1 and C_2, respectively, the locations of their Ho atoms are marked by yellow spheres, whereby the three possible Ho-Ho axes of C_2 cannot be distinguished.

A further species with symmetric STM contrast is encountered (orange outline), which distinguishes itself from C_2 through a brighter and more uniform appearance throughout the entire monomer. This species denoted C_3 is explained by the trapping of a third Ho atom within the chemical pocket of HEB. In consistency with the models for C_1 and C_2 (Fig. 5.11c), we propose the presence of three tweezer-like bonding motifs within C_3, which are arranged in a threefold symmetric fashion (only one of two possible scenarios is proposed by the C_3 model). While a trinuclear Co-HEB species with each Co atom confined between two neighbouring alkynes is absent, the existence of C_3 for Ho–HEB further indicates that the Ho atoms are located above the molecular plane for steric reasons. For comparison, the species C_1, C_2 and C_3 containing one, two and three π-like Ho-alkynyl motifs, respectively, are depicted in Fig. 5.11c by zoomed sections of STM images (hexagons) next to their proposed models. Considering the large diameter of Ho, we suspect that there are not more than three intramolecular binding sites available for this kind of organometallic complex.

The naturally formed isolated complexes within Ag(111) terraces are identified as C_3 units (Fig. 5.11d). To examine their structural integrity and flexibility on Ag(111), single-molecule manipulations are carried out on them. Figure 5.11d shows STM images before and after performing lateral manipulation via the STM tip. To this end, the tip is placed on the molecular center at a setpoint of $V_t = -10\,\text{mV}$, $I_t = 50\,\text{pA}$ before lowering it by 200 pm within the OFL mode. After guiding the complex along the trajectory of the moving STM tip (orange arrow), the subsequent STM image (bottom part of Fig. 5.11d) confirms the lateral displacement and structural integrity of the polynuclear complex upon tip-induced manipulation. Therefore, we assume that HEB and on-surface formed Ho–HEB (created at $T_{deposition} = 200$ K) represent the constituents of a mobile 2D gas that condensates into dense-packed domains during cooling down the sample for LT-STM imaging at ~4.5 K. Albeit the organoholmium phase and organic phase express the same domain orientations, we do not consider an immobile molecular layer during Ho deposition at 200 K (in contrast to in situ metallation at $T_{deposition} \leq 12$ K, see Fig. 5.3a, b), but assume the diffusion of both HEB and on-surface formed Ho–HEB at this temperature. The high molecular weight of C_3 (compared to C_2, C_1 and HEB) is supposed to lower its diffusion rate (higher barrier), thus rationalizing the natural occurence of isolated C_3 units next to Ho clusters in Fig. 5.11a (cf. isolated Co-HEB complexes in Fig. 5.7a).

The results indicate that the sixfold symmetric alkyne derivative HEB can serve as a host platform for the capture of multiple lanthanide atoms up to a cluster of three Ho atoms. By tuning the ionic diameter of the guest species, its bonding configuration and decoupling from Ag(111) can be controlled within a well-defined environment.

5.8 Conclusion

In summary, we have presented a systematic STM and DFT study of on-surface synthesized organometallic complexes, which were created by the incorporation of metal atoms into a multitopic alkyne derivative on Ag(111). Already in situ Co dosage at 12 K leads to the selective interaction of one Co atom with a pair of alkyne moieties, as predicted by gas-phase DFT calculations. By depositing Co at 200 K, the on-surface formation of a dicobalt compound is demonstrated. The organometallic building blocks self-assemble into a nanoarchitecture, whose embedded Co atoms express a 3.4.6.4 surface tessellation. Atom-precise STM imaging and extensive DFT calculations unravel the decisive role of the surface registry on the formation of the chiral tiling pattern, which is due to the site-selective adsorption behavior of the Co-HEB complex on Ag(111). A correlation between the complex's orientation and adsorption site is further corroborated by manipulation experiments on isolated complexes expressing discrete 120° rotations after tip-induced migrations to new adsorption sites. In addition, the STM tip-activated Co-HEB species expresses a Kondo feature serving as a chemical fingerprint for the incorporated Co atoms. Furthermore, manipulation experiments on isolated C_2 units demonstrate enantiomeric switching between their conformation states α and β, which are recognized within the enantiopure 2D domains of the organocobalt phase.

When using the 4f-block element Ho expressing a larger atomic diameter (compared to Co) for organometallation, site-selective metal-alkynyl bonding within HEB remain operative. However, since the organoholmium phase does not represent an ordered nanostructure, we assume the absence of site-selective interactions between the Ag(111) surface and embedded Ho atoms of Ho–HEB, which implies Ho adsorption on top of the molecular plane. While the two metal atoms of dinuclear Co-HEB are lying below the molecular plane, the dosage of Ho (instead of Co) generates polynuclear complexes embedding one, two and three Ho-alkyne motifs, each Ho atom situated above two adjacent alkynes.

Taking into account the extraordinary capabilities of the alkyne moiety to develop conjugated electronic states within a plethora of structures [16, 56–58], our results open up novel ways to generate nanoobjects with multiple spin-centers embedded within custom-designed molecular backbones. Such complexes could represent an alternative approach toward nanomagnetism, where wide tunability of the magnetic behavior is expected. Combining polynuclear alkynyl complexes with on-surface coupling reactions [59–61] could make possible the synthesis of extended spinsystems with prospects for quantum information manipulation at the single-molecule level. Our results bear fruitful perspectives for organometallic chemistry and heterogenous asymmetric catalysis, and provide useful prospectives toward the atom-precise fabrication of functional interfacial nanoarchitectures.

References

1. Brune H, Gambardella P (2009) Magnetism of individual atoms adsorbed on surfaces. Surf Sci 603:1812–1830
2. Enders A, Skomski R, Honolka J (2010) Magnetic surface nanostructures. J Phys Condens Matter 22:433001
3. Bogani L, Wernsdorfer W (2008) Molecular spintronics using single-molecule magnets. Nat Mater 7:179–186
4. N'Diaye AT, Bleikamp S, Feibelman PJ, Michely T (2006) Two-dimensional Ir cluster lattice on a graphene moiré on Ir(111). Phys Rev Lett 97:215501
5. Brihuega I, Michaelis CH, Zhang J, Bose S, Sessi V, Honolka J, Schneider MA, Enders A, Kern K (2008) Electronic decoupling and templating of co nanocluster arrays on the boron nitride nanomesh. Surf Sci 602:L95–L99
6. Pan Y, Gao M, Huang L, Liu F, Gao HJ (2009) Directed self-assembly of monodispersed platinum nanoclusters on graphene moiré template. Appl Phys Lett 95:093106
7. Natterer FD, Patthey F, Brune H (2012) Ring state for single transition metal atoms on boron nitride on Rh(111). Phys Rev Lett 109:066101
8. Bazarnik M, Brede J, Decker R, Wiesendanger R (2013) Tailoring molecular selfassembly of magnetic phthalocyanine molecules on Fe- and co-intercalated graphene. ACS Nano 7:11341–11349
9. Schlickum U, Decker R, Klappenberger F, Zoppellaro G, Klyatskaya S, Ruben M, Silanes I, Arnau A, Kern K, Brune H (2007) Metal-organic honeycomb nanomeshes with tunable cavity size. Nano Lett 7:3813–3817

10. Decker R, Schlickum U, Klappenberger F, Zoppellaro G, Klyatskaya S, Ruben M, Barth JV, Brune H (2008) Using metal-organic templates to steer the growth of Fe and Co nanoclusters. Appl Phys Lett 93:243102
11. Pivetta M, Pacchioni GE, Schlickum U, Barth JV, Brune H (2013) Formation of Fe cluster superlattice in a metal-organic quantum-box network. Phys Rev Lett 110:086102
12. Krenner W, Klappenberger F, Kühne D, Diller K, Qu ZR, Ruben M, Barth JV (2011) Positioning of single co atoms steered by a self-assembled organic molecular template. J Phys Chem Lett 2:1639–1645
13. Lang H, Weinmann M (1996) Bis(alkynyl) titanocenes as organometallic chelating ligands for the stabilization of monomeric organo copper(I) compounds. Synlett xx:1–10
14. Lang H, George DSA, Rheinwald G (2000) Bis(alkynyl) transition metal complexes, $R^1C{\equiv}C$-[M]-$C{\equiv}CR^2$, as organometallic chelating ligands; formation of $\mu,\eta^{1(2)}$-alkynylbridged binuclear and oligonuclear complexes. Coord Chem Rev 206:101–197
15. Powell CE, Humphrey MG (2004) Nonlinear optical properties of transition metal acetylides and their derivatives. Coord Chem Rev 248:725–756
16. Costuas K, Rigaut S (2011) Polynuclear carbon-rich organometallic complexes: clarification of the role of the bridging ligand in the redox properties. Dalton Trans 40:5643–5658
17. Wong WY, Ho CL (2010) Organometallic photovoltaics: a new and versatile approach for harvesting solar energy using conjugated polymetallaynes. Acc Chem Res 43:1246–1256
18. Whittell GR, Hager MD, Schubert US, Manners I (2011) Functional soft materials from metallopolymers and metallosupramolecular polymers. Nat Mater 10:176–188
19. Čechal J, Kley CS, Kumagai T, Schramm F, Ruben M, Stepanow S, Kern K (2013) Functionalization of open two-dimensional metal-organic templates through the selective incorporation of metal atoms. J Phys Chem C 117:8871–8877
20. Weber PB, Hellwig R, Paintner T, Lattelais M, Paszkiewicz M, Casado Aguilar P, Deimel PS, Guo Y, Zhang Y-Q, Allegretti F (2016) Surface-guided formation of an organocobalt complex. Angew Chem Int Ed 55:5754–5759
21. Omae I (2007) Three characteristic reactions of organocobalt compounds in organic synthesis. Appl Organomet Chem 21:318–344
22. Ruben M Institute of Nanotechnology, Karlsruhe Institute of Technology (KIT), 76344 Eggenstein-Leopoldshafen (DE), https://www.int.kit.edu/ruben.php
23. Kovács A, Frenking G (1999) Bonding interactions of a molecular pair of tweezers with transition metals: Theoretical study of bis(η^2-alkyne) complexes of copper(I), silver(I), and gold(I)[1]. Organometallics 18:887–894
24. Lang H, Herres M, Imhof W (1994) Stabiliziation of low-coordinated M(CO) buildingblocks: synthesis and reactivity of $[(\eta^5\text{-}C_5H_4SiMe_3)_2\ Hf(C{\equiv}CPh)_2]$ M(CO) (M=Ni, Co). J Organomet Chem 465:283–287
25. Herres M, Lang H (1994) Stable (η^2-alkyne) MCl_2 complexes–structure and reactivity. J Organomet Chem 480:235–239
26. Kühne D, Klappenberger F, Decker R, Schlickum U, Brune H, Klyatskaya S, Ruben M, Barth JV (2009) High-quality 2d metal-organic coordination network providing giant cavities within mesoscale domains. J Am Chem Soc 131:3881–3883
27. Chavey D (1989) Tilings by regular polygons–II: a catalog of tilings. Comput Math Appl 17:147–165
28. Kepler J (1619) Harmonices mundi, Johannes Planck, Linz
29. Schlickum U, Decker R, Klappenberger F, Zoppellaro G, Klyatskaya S, Auwärter W, Neppl S, Kern K, Brune H, Ruben M, Barth JV (2008) Chiral kagomé lattice from simple ditopic molecular bricks. J Am Chem Soc 130:11778–11782
30. Écija D, Urgel JI, Papageorgiou AC, Joshi S, Auwärter W, Seitsonen AP, Klyatskaya S, Ruben M, Fischer S, Vijayaraghavan S, Reichert J, Barth JV (2013) Fivevertex archimedean surface tessellation by lanthanide-directed molecular self-assembly. Proc Natl Acad Sci USA 110:6678–6681
31. Stepanenko V, Kandanelli R, Uemura S, Wurthner F, Fernandez G (2015) Concentration-dependent rhombitrihexagonal tiling patterns at the liquid/solid interface. Chem Sci 6:5853–5858

32. Seitsonen Ari P École Normale Supérieure - Chemistry Department, 24 rue Lhomond, F-75231 Paris Cedex 05 - France, http://www.chimie.ens.fr/?q=en/pasteur/pct/Ari_Paavo_Seitsonen/
33. Tseng T-C, Urban C, Wang Y, Otero R, Tait SL, Alcamí M, Écija D, Trelka M, Gallego JM, Lin N (2010) Charge-transfer-induced structural rearrangements at both sides of organic/metal interfaces. Nat Chem 2:374–379
34. Zhang L, Bagrets A, Xenioti D, Korytár R, Schackert M, Miyamachi T, Schramm F, Fuhr O, Chandrasekar R, Alouani M, Ruben M, Wulfhekel W, Evers F (2015) Kondo effect in binuclear metal-organic complexes with weakly interacting spins. Phys Rev B 91:195424
35. DiLullo A, Chang S-H, Baadji N, Clark K, Klöckner J-P, Prosenc M-H, Sanvito S, Wiesendanger R, Hoffmann G, Hla S-W (2012) Molecular kondo chain. Nano Lett 12:3174–3179
36. Luis F, Repollés A, Martínez-Pérez MJ, Aguilá D, Roubeau O, Zueco D, Alonso PJ, Evangelisti M, Camón A, Sesé J, Barrios LA, Aromí G (2011) Molecular prototypes for spin-based CNOT and SWAP quantum gates. Phys Rev Lett 107:117203
37. Aromi G, Aguila D, Gamez P, Luis F, Roubeau O (2012) Design of magnetic coordination complexes for quantum computing. Chem Soc Rev 41:537–546
38. Aguilà D, Barrios LA, Velasco V, Roubeau O, Repollés A, Alonso PJ, Sesé J, Teat SJ, Luis F, Aromí G (2014) Heterodimetallic [LnLn'] lanthanide complexes: toward a chemical design of two-qubit molecular spin quantum gates. J Am Chem Soc 136:14215–14222
39. Westerström R, Dreiser J, Piamonteze C, Muntwiler M, Weyeneth S, Brune H, Rusponi S, Nolting F, Popov A, Yang S, Dunsch L, Greber T (2012) An endohedral single-molecule magnet with long relaxation times: DySc2N@C80. J Am Chem Soc 134:9840–9843
40. Zadrozny JM, Xiao DJ, Atanasov M, Long GJ, Grandjean F, Neese F, Long JR (2013) Magnetic blocking in a linear iron(I) complex. Nat Chem 5:577–581
41. Ungur L, Le Roy JJ, Korobkov I, Murugesu M, Chibotaru LF (2014) Fine-tuning the local symmetry to attain record blocking temperature and magnetic remanence in a single-ion magnet. Angew Chem Int Ed 53:4413–4417
42. Wäckerlin C, Donati F, Singha A, Baltic R, Rusponi S, Diller K, Patthey F, Pivetta M, Lan Y, Klyatskaya S, Ruben M, Brune H, Dreiser J (2016) Giant hysteresis of singlemolecule magnets adsorbed on a nonmagnetic insulator. Adv Mater 28:5195–5199
43. Kouwenhoven L, Glazman L (2001) Revival of the kondo effect. Phys World 14:33
44. Nagaoka K, Jamneala T, Grobis M, Crommie MF (2002) Temperature dependence of a single kondo impurity. Phys Rev Lett 88:077205
45. Kröger J, Limot L, Jensen H, Berndt R, Crampin S, Pehlke E (2005) Surface state electron dynamics of clean and adsorbate-covered metal surfaces studied with the scanning tunnelling microscope. Prog Surf Sci 80:26–48
46. Schneider MA, Wahl P, Diekhöner L, Vitali L, Wittich G, Kern K (2005) Kondo effect of Co adatoms on Ag monolayers on noble metal surfaces. Jpn J Appl Phys 44:5328
47. Iancu V, Hla S-W (2006) Realization of a four-step molecular switch in scanning tunneling microscope manipulation of single chlorophyll-a molecules. Proc Natl Acad Sci USA 103:13718–13721
48. Tierney HL, Murphy CJ, Jewell AD, Baber AE, Iski EV, Khodaverdian HY, McGuire AF, Klebanov N, Sykes ECH (2011) Experimental demonstration of a single-molecule electric motor. Nat Nanotechnol 6:625–629
49. Auwärter W, Seufert K, Bischoff F, Ecija D, Vijayaraghavan S, Joshi S, Klappenberger F, Samudrala N, Barth JV (2012) A surface-anchored molecular four-level conductance switch based on single proton transfer. Nat Nanotechnol 7:41–46
50. Perera UGE, Ample F, Kersell H, Zhang Y, Vives G, Echeverria J, Grisolia M, Rapenne G, Joachim C, Hla S-W (2013) Controlled clockwise and anticlockwise rotational switching of a molecular motor. Nat Nanotechnol 8:46–51
51. Alemani M, Peters MV, Hecht S, Rieder KH, Moresco F, Grill L (2006) Electric field-induced isomerization of azobenzene by STM. J Am Chem Soc 128:14446–14447
52. Zhang JL, Xu JL, Niu TC, Lu YH, Liu L, Chen W (2014) Reversible switching of a single-dipole molecule imbedded in two-dimensional hydrogen-bonded binary molecular networks. J Phys Chem C 118:1712–1718

53. Parschau M, Passerone D, Rieder KH, Hug H, Ernst KH (2009) Switching the chirality of single adsorbate complexes. Angew Chem Int Ed 48:4065–4068
54. Shannon RD (1976) Revised effective ionic radii and systematic studies of interatomic distances in halides and chalcogenides. Acta Crystallogr Sect A Cryst Phys Diffr Theor Gen Crystallogr 32:751–767
55. Lide DR (2004) Handbook of chemistry and physics. CRC Press LLC
56. Nielsen MB, Diederich F (2005) Conjugated oligoenynes based on the diethynylethene unit. Chem Rev 105:1837–1867
57. Liu J, Lam JWY, Tang BZ (2009) Acetylenic polymers: syntheses, structures, and functions. Chem Rev 109:5799–5867
58. Zhang Y-Q, Björk J, Barth JV, Klappenberger F (2016) Intermolecular hybridization creating nanopore orbital in a supramolecular hydrocarbon sheet. Nano Lett 16:4274–4281
59. Klappenberger F, Zhang YQ, Björk J, Klyatskaya S, Ruben M, Barth JV (2015) Onsurface synthesis of carbon-based scaffolds and nanomaterials using terminal alkynes. Acc Chem Res 48:2140–2150
60. Dong L, Liu PN, Lin N (2015) Surface-activated coupling reactions confined on a surface. Acc Chem Res 48:2765–2774
61. Fan QT, Gottfried JM, Zhu JF (2015) Surface-catalyzed C-C covalent coupling strategies toward the synthesis of low-dimensional carbon-based nanostructures. Acc Chem Res 48:2484–2494

Chapter 6
Ho-Catalyzed Cyclotrimerization

The author of this thesis published the results of this chapter along with additional insights from recent experiments in the journal "Chem. Eur. J." (see chapter "List of publications").

6.1 Introduction

The on-surface coupling of terminal alkynes on different metal surfaces has provided new synthesis protocols for the creation of low-dimensional carbon structures (see Sect. 1.3.5). The Ag(111) surface strongly favors the homocoupling reaction pathway resulting in chain-like butadiyne linking motifs after hydrogen abstraction (see Fig. 4.1b) [1, 2]. On Au(111), however, the coupling of three terminal alkynes toward a substituted benzene compound is chemically favored and realized in a two-stage coupling process [3]. Theory calculations on reaction pathways predict for both reaction types an intermediate stage, where organometallic metal-carbon bonds with Ag/Au atoms catalyze intermolecular coupling through the conversion from triple to double bonds (see coupling steps in Fig. 4.1b) [4, 5]. In the context of the tritopic alkyne molecule Ext-TEB, mesoporous polymer structures were created through heat-induced homocoupling and cyclotrimerization (see Fig. 1.5e–k). The structurally similar molecule TEB deposited on the highly reactive Rh(111) substrate facilitates the on-surface synthesis of graphene islands [6]. Ext-TEB on Cu(111) results in deprotonated monomers that are stabilized by a trifurcated ionic hydrogen bond [7]. A different alkyne-derivative (1,4-diethynylbenzene) on Cu(111) shows both homocoupling and cyclotrimerization besides other side reactions, but with apparently high reaction barriers and very low chemoselectivity [8]. An uncertainty in unraveling the respective reaction route for terminal alkyne coupling arises from the occurrence of organometallic bonds with substrate atoms, which are interpreted as weak intermediates on Ag(111)/Cu(111)/Au(111), but represent strong

R. Hellwig, *Alkyne-Based Nanostructures on Silver Substrates*, Springer Theses,
https://doi.org/10.1007/978-3-030-00997-7_6

Ag-bis-acetylide linkage on the Ag(110) surface (see Chap. 3 and Ref. [9]). Depending on the chemical character of the substrate, the substrate atoms are supposed to catalyze the coupling reaction.

An alternative approach toward on-surface reactions is based on the depositon of catalytic metal atoms (heteroatoms) in order to catalyze coupling reactions, e.g., Pd- and Cu-catalyzed Ullmann coupling (see Sect. 1.3.5) on Au(111) [10] and on a hexagonal boron nitride layer supported on Ni(111) [11]. On weakly interacting surfaces, the reaction mechanism depends on the chemical aspects of the employed catalytic species (cf. solution-based coupling reactions in Fig. 1.2b). In search of an adequate catalytic agent for coupling reactions with Ext-TEB monomers on Ag(111), we initially tested the transition elements Co and Pd, which are well known from cycloaddition reactions in solution-based environments [12]. Related on-surface reactions were reported by Abbet et al., who utilized thermal desorption and Fourier transform infrared spectroscopy to study Pd-catalyzed acetylene cyclotrimerization on MgO films [13]. However, within the targeted temperature regime, i.e., below the threshold for surface-catalyzed homocoupling, neither Pd nor Co catalyzed covalent bond formation between Ext-TEB monomers on Ag(111): Pd mediates the formation of organometallic Pd-bis-acetylide chains, while Co does not trigger on-surface linkage.

Here, we present a lanthanide-catalyzed dimerization and cyclotrimerization reaction requiring terminal alkyne molecules and single Ho atoms on a Ag(111) surface. Since the reaction takes place at low temperatures ($\sim$200 K), it exhibits clear advantages over other coupling schemes requiring thermal activation above room temperature, at which detrimental side reactions and dissociation products (e.g., halogens) as well as monomer desorption may occurr.

Ho belongs to the 4f-block elements with atomic numbers ranging from 57 to 71, which express the electronic ground state configurations [Xe]$4f^n6s^2$ and [Xe]$4f^n5d^16s^2$. Their chemical properties are related to the presence of 4f electrons, which are protected by surrounding $5s^2$ and $5p^6$ orbitals from interactions with their environment [14]. While crystal field interactions play an important role in transition metal-based metal-organic coordination networks (MOCN), lanthanide-based metal-ligand interactions are often dominated by electrostatic and steric effects at coordination centers [15, 16]. Similar to alkali and alkaline earth elements, lanthanides express electropositive oxidation states, which mediate flexible ionic bonding geometries, mostly stabilized by a trivalent ion with an inert closed-shell electron configuration. Their atomic diameter varies over the 4f series according to the so-called lanthanide contraction, i.e., the 4f electrons cannot effectively shield the increase in nuclear charge, which causes the decrease of the atomic diameter with increasing atomic number [16]. The large ionic diameter of 4f-block elements supports high-coordination bonding scenarios with electronegative ligands, thereby offering new routes toward metal-organic engineering, e.g., the on-surface creation of sandwich complexes [17], fivefold lanthanide-ligand coordination networks expressing Archimedian tesselation [18] and metal-organic coordination networks (MOCN) with high thermal stability [19]. Besides their high potential for molecular magnetism [20, 21], surface-supported lanthanide atoms reveal exciting magnetic properties due to a high magnetic moment and strong magnetoelastic anisotropy stemming from spin-orbit coupling [22]. Recently, single-atom magnets with long relaxation times and magnetic remanence up to a temperature of 30 K were reported [23]. Furthermore, lanthanide atoms exhibit unique luminescence characteristics arising from 4f to 4f transitions [24].

The 4f-block elements and related organolanthanide complexes are also interesting building blocks for organometallic chemistry and organic synthesis. In the last chapter, we reported on the on-surface formation of lanthanide alkynyl complexes expressing site-selective organometallic π bonding with the multiple internal alkynes of HEB (see Sect. 5.7). Solution-based reactions between lanthanide atoms and alkyne derivatives resulted in the reduction of alkynes and alkenes [25]. Furthermore, it was reported that organolanthanide complexes can catalyze terminal alkyne dimerization in solution [26]. In this context, it would be interesting to explore by means of LT-STM the catalytic properties of surface-supported lanthanide atoms in activating on-surface reactions with terminal alkynes.

6.2 Ho-Catalyzed Dimerization Reaction

In the following, we will employ Ho atoms to catalyze dimerization and cyclotrimerization reactions with terminal alkyne monomers confined on Ag(111). First, a sub--ML amount of Ext-TEB ($T_{sublimation}$ = 420 K) was deposited on the clean Ag(111) sample at temperatures below 250 K, in order to avoid homocoupling processes proceeding at $T_{substrate} \geq 330$ K [1]. Sample temperatures far below RT were established through permanent cooling of the manipulator with liquid nitrogen and, if necessary, slow and controlled counter-heating of the sample. For the precise control of $T_{substrate}$ during sample preparation, the thermocouple only contacts the Ag(111) crystal (sidewise). Prior to the deposition of the catalytic agent, the prepared sample was cooled down to liquid He temperatures and characterized by STM. As depicted in Fig. 6.1a, the molecules establish a chiral organization within extended dense-packed arrays of nanopores, each stabilized through weak C–H$\cdots\pi$ hydrogen bonding. In detail, each nanopore is formed by the cyclic arrangement of six pristine alkyne moieties, whose intermolecular interactions give rise to a delocalized supramolecular orbital [27]. Empty Ag(111) terraces and step edges are decorated by single molecules imaged as triangular-shaped protrusions (inset). After confirming the absence of organosilver structures and homocoupling, Ho is deposited on the sample kept at 130 K. This sample temperature is maintained for 30 min (min) in order to ensure the completion of Ho-induced reactions at low temperatures. Resuming STM measurements at 4.4 K show the dissolution of the pristine Ext-HEB phase (Fig. 6.1b), while small areas of the molecular adlayer bespeak the pristine arrangement of the organic phase. Laterally displaced and liberated monomers are encountered at the regions close to island edges and close to scattered bright blobs, which are attributed to Ho clusters decorating molecular islands and Ag(111) terraces. In the magnified STM image of Fig. 6.1c, one sees that all Ho clusters are surrounded by monomers. Furthermore,

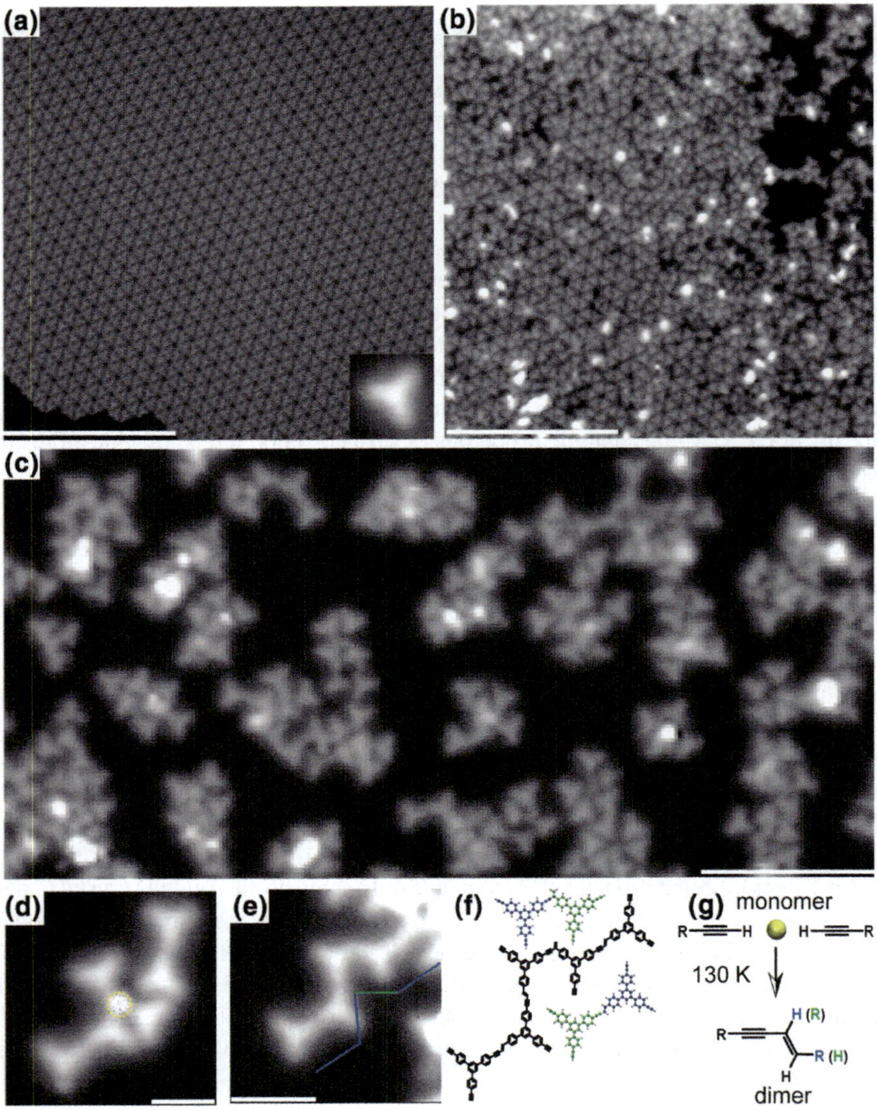

Fig. 6.1 STM results for Ho dosage at 130 K. **a** Pristine phase of Ext-TEB on Ag(111). Inset shows a single monomer. **b** Sample after Ho deposition at 130 K. **c** STM topograph shows dissolution of Islands and formation of covalent structures. **d** Dimers next to a Ho cluster (yellow dotted circle). **e** STM image showing a covalent oligomer. **f** Corresponding structure model along with two dimer models for enyne regioisomers (blue and green line in **e**). **g** First step of Ho-assisted reaction pathway generating two dimeric regioisomers according to coloring of hydrogen atoms (H) and residue moieties (R). Tunneling parameters V_t, I_t: **a** 1 V, 0.3 nA (inset: 0.5 V, 0.3 nA); **b**–**e** 0.5 V, 0.1 nA. Scale bars: **a, b** 20 nm (inset of **a**: 2.48 nm); **c** 10 nm; **d** 2 nm; **e** 3 nm. Coloring: H (white), C (green and blue) and Ho (yellow)

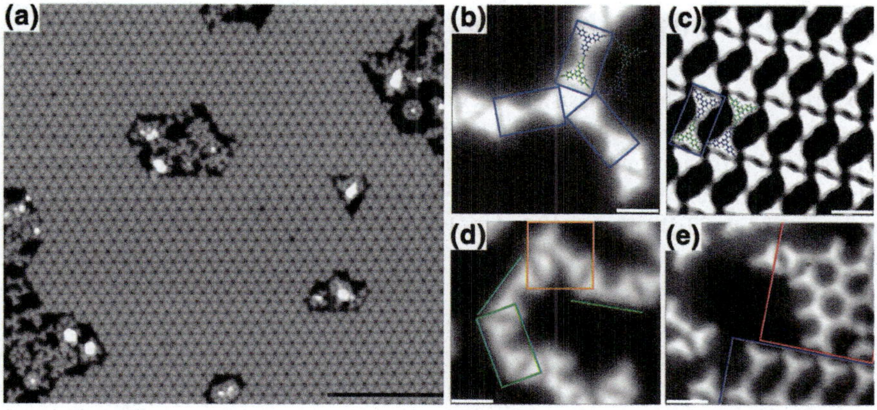

Fig. 6.2 STM results for Ho dosage at 200 K. **a** Large-scale STM image showing dissolution of organic phase after Ho dosage. **b** Three covalent dimers (blue outline) surrounding one monomer. **c** Enantiopure assembly of dimer species (first regioisomer). **d** Second regioisomer of dimerization (green outline) and asymmetric trimer species (orange outline). **e** Covalent oligomer structure comprised of threefold coupling nodes (red outline). Tunneling parameters V_t, I_t: **a** 0.3 V, 0.3 nA; **b** 0.5 V, 0.1 nA; **c** 0.05 V, 0.3 nA; **d** 0.3 V, 0.3 nA; **e** 0.05 V, 0.3 nA. Scale bars: **a** 20 nm; **b–e** 2 nm

dimer and oligomer structures not observed before Ho deposition appear. Specifically, Fig. 6.1d shows two dimers located next to a Ho cluster, which is marked by a yellow dotted circle. A closer look at the coupling motif reveals that the two interconnected monomer end groups are slightly offset from each other due to a kink along the dimer axis (more clearly visible in Fig. 6.2c). This is a strikingly different bonding scenario compared to terminal alkyne homocoupling resulting in a straight dimer axis (cf. Figs. 3b, c and 5b of Ref. [1]). Accordingly, we explain the kink motif of the dimer by the conversion of one triple bond into a double bond (see structure model in Fig. 6.1g). Besides single dimers, also covalent oligomers are encountered, as shown in Fig. 6.1e. The oligomer structure comprised of five molecules contains a further twofold coupling motif with a kink (green line). From the corresponding oligomer model and dimer models in Fig. 6.1f, it follows that both species belong to the class of enynes. Equivalent enyne products were reported by Nishiura et al., who used lanthanide half-metallocene complexes to catalyze terminal alkyne dimerization in solution environment [26].

According to Fig. 6.1d displaying a Ho cluster surrounded by monomers and dimers, the coupling scheme proposed in Fig. 6.1g refers to a Ho-catalyzed reaction operative at temperatures down to T $\sim$ 130 K; the two regioisomers are recognized by the positions of color-coded hydrogen atoms (H) and residue moieties (R). While a Ho-catalyzed ethynyl→vinyl conversion of a single monomer would require two extra hydrogen atoms, vinyl-ethynyl catenation proceeds without the uptake or release of hydrogen atoms.

6.3 Ho-Catalyzed Cyclotrimerization Reaction

To further explore the Ho-assisted reaction route, we investigated the scenario of Ho dosage at 200 K. The higher sample temperature and associated increase in thermal energy and diffusion rates of the adsorbates is supposed to permit the activation of states with higher reaction barriers. After dosing minute Ho amounts on a sub-ML adlayer of Ext-TEB, the gradual dissolvement of organic islands is observed. In contrast to the preparation at 130 K (cf. Fig. 6.1b), the STM topograph in Fig. 6.2a reveals that Ho clusters do not decorate self-assembled Ext-HEB islands anymore, but are well separated from dense-packed monomer regions. Interestingly, they are surrounded by regions of free space, whereby monomers detaching from nearby island edges can interact with Ho. The observed coupling motifs within these regions further consolidates our interpretation of Ho-catalyzed on-surface coupling. Since the Ho clusters next to their catalyzed structures are enclosed by dense-packed assemblies of pristine Ext-TEB monomers (Fig. 6.2a), we conclude that the reaction yield can be controlled by the Ho concentration.

The preceding STM results for Ho deposition at 130 K show the presence of Ho clusters (cf. Fig. 6.1b), thus indicating a considerable adsorbate mobility at this temperature. During the diffusion of Ho atoms, attractive forces and aggregation of Ho atoms may give rise to continuous cluster growth. Owing to the decrease in cluster mobility with increasing size, there is a critical size that inhibits further diffusion. At this stage, pristine molecules are supposed to migrate to the cluster and undergo coupling reactions (cf. Fig. 6.2a). Since the catalytic properties of a cluster may alter dramatically with its size, the reaction might stop at a certain cluster size. Furthermore, it is not clear whether single-atom Ho centers suffice to trigger coupling reactions.

The STM image in Fig. 6.2b shows three isochiral dimers (blue outline) around one monomer. In consistence with the preceding results (cf. Fig. 6.1d and e for preparation at 130 K), one recognizes a kink along the dimer axis indicative of vinyl-ethynyl catenation. Since we cannot state which side of the dimeric enantiomer underwent the ethynyl→vinyl conversion, two gas-phase molecular models (simulated by the Hyperchem software [28]) with opposite orientations are superimposed. Besides isolated dimer patterns, we also encountered an enantiopure assembly of on-surface formed dimers (Fig. 6.2c). Therein, the dimers express the opposite handedness compared to the dimers in Fig. 6.2b. The second enyne species from Fig. 6.1e is again scarcely observed after Ho-induced reactions at 200 K (green outline in Fig. 6.2d). In contrast to Ho dosage at 130 K, we identify threefold coupling motifs within isolated trimer species (orange outline of Fig. 6.2d) and extended oligomer structures (red outline in Fig. 6.2e).

As demonstrated by the STM topograph in Fig. 6.3a, there are four coupling motifs in total, i.e., two dimeric regioisomers expressing vinyl-ethynyl catenation (blue and green outlines) and two trimeric regioisomers formed through cyclotrimerization (orange and red outlines). The adjoining structure models in Fig. 6.3b reveal color-scaled monomers that are coupled via the coupling of three terminal alkynes into

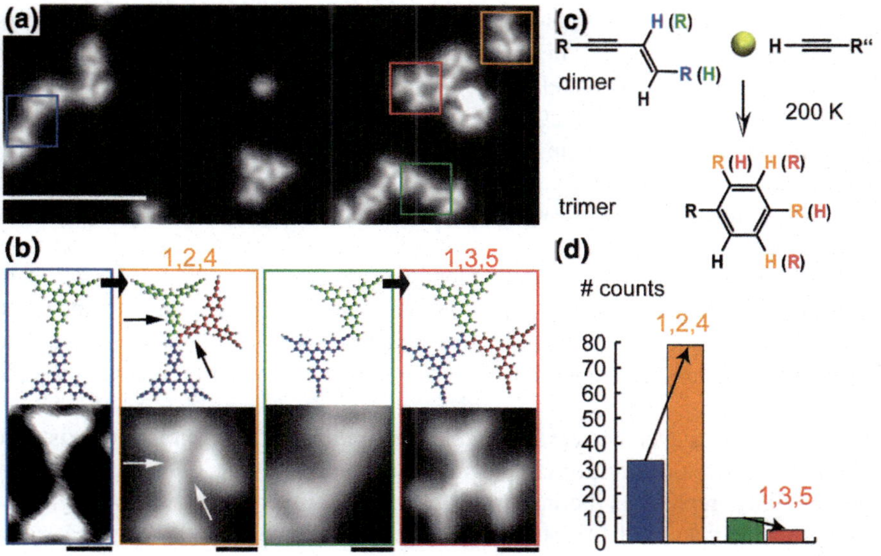

Fig. 6.3 **a** Coexistence of dimerization and cyclotrimerization (colored rectangles) within STM image. **b** High-resolution images of dimerization motifs (blue and green frames) and cyclotrimerization motifs (orange and red frames). Adjoined gas-phase models created via Hyperchem software [28]. **c** Second step of Ho-assisted cyclotrimerization. Color-coding of H and R to distinguish regioisomers for both the intermediates and products. **d** Statistical counting of intermediate and product species within STM image (~434 molecules within area of 100 × 100 nm^2). Tunneling parameters V_t, I_t: **a** 0.5 V, 0.1 nA; **b** from left to right: 0.05 V, 0.3 nA; 0.5 V, 0.1 nA; 0.3 V, 0.3 nA; 0.05 V, 0.3 nA. Scale bars: **a** 10 nm; **b** 1 nm. Coloring: H (white), C (red, green and blue) and Ho (yellow)

an asymmetric 1,2,4- and symmetric 1,3,5-trisubstituted benzene ring. While both dimeric regioisomers are chiral, only the 1,2,4-cyclotrimer is chiral. Without STM image simulations, one might guess that its two nodal planes (arrows) within the STM contrast (orange outline) are related to non-covalent bonding or Ho-triggered molecular fragmentation. However, from our observation that all asymmetric trimers are congruent and always express the same STM imaging characteristics (locations of nodal planes with respect to long axis of trimer), together with the good agreement with the adjoined gas-phase model (above), we identify this species as the reaction product of 1,2,4-cyclotrimerization. The two depressions (nodal planes within STM contrast) of the trimer coincide with two adjacent phenyl rings, thus indicating that steric and electrostatic repulsion between C–H moieties is operative. The latter supposedly causes a non-planar alignment (rotation) of the respective two phenyl compounds, thus leading to a local decrease of the tunneling conductance (possibly due to neglectable overlap between π orbitals and tip states). In case of cyclotrimerization of diyne linkers on Au(111), Zhou et al. explained the low amount of asymmetric cyclotrimers on Au(111) by unfavorable interfacial interactions compared to the planar configuration of symmetric cyclotrimers [4]. Using the monomer Ext-TEB for cyclotrimerization on Au(111), Liu et al. reported the exclusive formation of 1,3,5-cyclotrimerization [3]. In our case, the predominance of asymmetric reaction products on Ag(111) may be related to the size and shape distribution of Ho clusters formed during Ho dosage at 200 K. Possibly, single-atom Ho centers favor a different ratio between both regioisomeric forms than large Ho clusters (cf. Fig. 6.2a) exhibiting steric constrains for the coupling reaction. The assumed rotation of the two phenyl rings (causing the nodal planes in the STM contrast) of the asymmetric trimer (Fig. 6.3b) may already take place before Ho-induced C–C bond formation. This would mean that a dimer and a monomer next to a Ho cluster (two-step coupling process explained below) rotate the pertaining phenyl groups in response to steric constrains, which may be due to interactions between molecules, or due to site-selective organolanthanide bonds during ethynyl → vinyl reduction at the Ho cluster.

The absence of cyclotrimers for sample preparation at 130 K, and the coexistence of dimers and cyclotrimers for sample preparation at 200 K (Fig. 6.3a and b) suggests that the dimer represents the intermediate for cyclotrimerization. Accordingly, we propose a Ho-catalyzed two-step reaction scheme involving two coupling steps. The first step of this reaction is already present at substrate temperatures down to ~130 K, as schematically proposed in Fig. 6.1g. The second step originating from Ho-catalyzed coupling between one dimer and one pristine monomer is schematically depicted in Fig. 6.3c. The chosen colors of H and R moieties (cf. Figs. 6.1e, g and 6.3a, c) differentiate the two regioisomeric forms for twofold and threefold coupling, respectively. Our proposed two-step model agrees with previously reported studies, e.g., acetylene cyclotrimerization on Pd(111) via a C_4H_4 intermediate [29, 30]. This intermediate also appears for the acetylene→benzene cyclotrimerization on Cu(110), whose reaction yield is limited by benzene desorption [31, 32].

To quantitatively compare the distribution of coupling motifs, a STM image with ~434 molecules ($100 \times 100\,\text{nm}^2$ size) was statistically evaluated. One fourth of all terminal alkynes are involved in covalent linking motifs, i.e., twofold vinyl-ethynyl catenation and threefold connected benzene compounds, as depicted in the histogram below (Fig. 6.3d). The ratio between twofold (vinyl-ethynyl linkage) and threefold (cylclotrimerization) coupling nodes is approximately 1:2. On the intermediate side of the diagram (blue and green bars), a ratio of ~3:1 indicates that the dimer motif within the blue frame is strongly favored over the other regioisomer (green frame). On the product side (orange and red), a striking ratio of ~16:1 implies the strong dominance of asymmetric coupling motifs over symmetric 1,3,5-trisubstituted benzene compounds. Based on comparing these ratios, we tentatively speculate that asymmetric 1,2,4-cyclotrimers emerge from the frequently occurring intermediate species (blue frame in Fig. 6.3b), while the infrequent dimer species (green frame) converts into symmetric 1,3,5-cyclotrimers (see arrows in Fig. 6.3d). We did not observe additional reaction products besides 1,2,4,- and 1,3,5-cyclotrimerization motifs, i.e., fourfold coupling nodes related to alkyne cyclotetramerization are absent, thus implying a strong chemoselectivity for this Ho-induced coupling reaction on Ag(111).

An increase in the Ho concentration would further shift the statistical weight on the side of cyclotrimerization, thereby reducing the amount of monomers and twofold

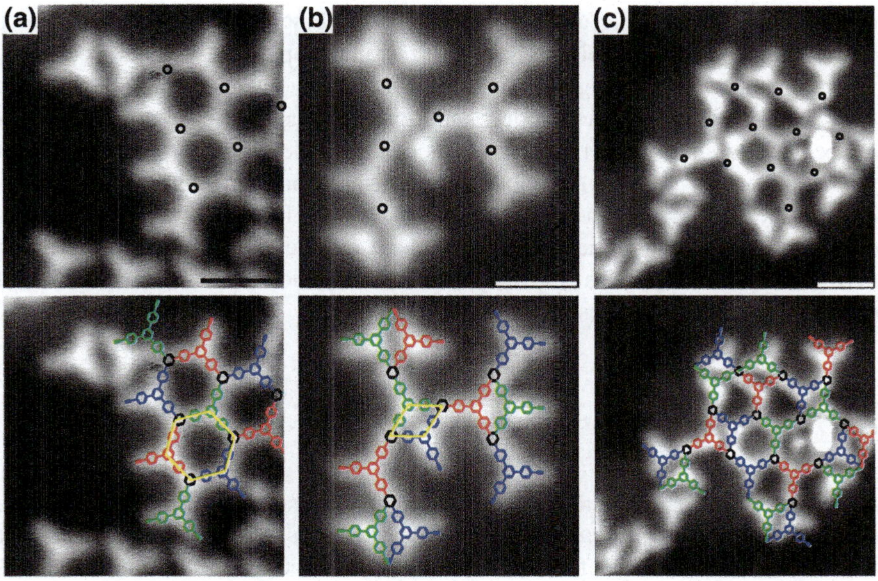

Fig. 6.4 Covalent oligomer structures exclusively comprised of threefold coupling nodes. **a** Regular honeycomb-like scaffold formed via 1,3,5-cyclotrimerization. **b** Non-regular pattern resulting from 1,2,4-cyclotrimerization. **c** Mixed hydrocarbon scaffold containing symmetric and asymmetric coupling nodes. Black circles depict coupling nodes. Bottom part of **a**–**c**: Superimposed structure models (Chemdraw software [33]) with color-coded molecules, whereby on-surface formed substituted benzene compounds are black. Yellow outlines emphasize different pore shapes: hexagon and rhomboid. Tunneling parameters V_t, I_t: **a** 0.05 V, 0.3 nA; **b** 0.5 V, 0.1 nA; **c** 0.05 V, 0.3 nA. Scale bars: **a**–**c** 20 nm

coupling motifs. At a certain level at which no monomers are left due to complete conversion into dimeric structures, the latter cannot be further transformed via the second reaction step. This implies that polymer structures generated via exceeding Ho-monomer ratios can contain residual twofold linkage nodes, in agreement with our experimental findings. Since cyclotrimerization sets in at sample temperatures down to ~185 K and at catalytic Ho amounts, we conclude that this 4f-block element is an efficient catalytic agent with remarkably low activation barriers for the on-surface coupling on inert surfaces.

When sufficiently increasing the Ho amount, we observe the predominance of open-porous carbon scaffolds instead of isolated cyclotrimers. Figure 6.4 displays STM images of extended mesoporous nanoarchitectures containing symmetric or/and asymmetric linkage nodes (black circles). To illustrate the polymer structures, each STM image is displayed together with an adjoining structure model (bottom) with colored monomers. Within these models (Chemdraw software [33]), black-colored benzene compounds indicate cyclotrimerization nodes. A regular oligomeric patch of 1,3,5-trisubstituted motifs is depicted in Fig. 6.4a. More frequently are non-regular aggregates formed via the 1,2,4-cyclotrimerization of terminal alkynes (Fig. 6.4b). In addition, there are mixed oligomeric networks incorporating both regioisomeric motifs (Fig. 6.4c). As shown in the bottom part of Fig. 6.4a and b (see yellow outlines), the different pore sizes and shapes are ascribed to primary void elements, i.e., a hexagon-shaped pore framed by six molecules establishing three 1,3,5-cyclotrimerization nodes, and a rhomboid-shaped pore formed by four molecules establishing two 1,2,4-cyclotrimerization nodes. From the good agreement with the superimposed polymer models, we conclude that all asymmetric motifs (cf. nodal planes in Fig. 6.3b) in Fig. 6.4b and c result from 1,2,4-cylotrimerization, thus excluding the presence of assembled molecular fragments within the hydrocarbon networks. Accordingly, this Ho-catalyzed alkyne reaction leads to the formation of extended covalent structures on the Ag(111) surface.

6.4 Conclusion

In summary, we have demonstrated the potential of surface-adsorbed lanthanide atoms in catalyzing terminal alkyne coupling reactions at low temperatures. After the successive deposition of the terminal alkyne molecule Ext-TEB and Ho atoms on Ag(111), we encounter the dissolution of organic islands and formation of covalent structures near Ho clusters. The Ho-induced reaction is well separated from surface-catalyzed homocoupling starting at higher sample temperatures on Ag(111) ($T \geq 330$ K). For Ho dosage at 130 K, we exclusively observe twofold coupling motifs within dimers and chains, indicating intermolecular linkage through enyne bridges. For sample preparations with Ho dosed at 200 K, the analysis of coexisting twofold and threefold coupling motifs implies a two-step cylotrimerization reaction, i.e., dimerization of Ext-TEB precedes cyclotrimerization. In contrast to the surface-catalyzed cyclotrimerization of Ext-TEB on Au(111), asymmetric coupling nodes (1,2,4-trisubstituted benzene rings) dominate over symmetric reaction products on Ag(111). The STM contrast of the asymmetric cyclotrimer is assigned to the non-planarity (rotations) of two phenyl rings due to steric reasons. Based on the observed open-porous nanomeshes expressing symmetric and asymmetric coupling nodes, we conclude that this low-temperature reaction is well suited for the on-surface polymerization toward covalent organic frameworks. The Ho-induced linkage of (small) molecules on weakly interacting surfaces may become interesting for applications in the field of molecular (opto)electronics [34] and heterogenous catalysis.

References

1. Zhang Y-Q, Kepčija N, Kleinschrodt M, Diller K, Fischer S, Papageorgiou AC, Allegretti F, Björk J, Klyatskaya S, Klappenberger F, Ruben M, Barth JV (2012) Homocoupling of terminal alkynes on a noble metal surface. Nat Commun 3:1286
2. Björk J, Zhang Y-Q, Klappenberger F, Barth JV, Stafström S (2014) Unraveling the mechanism of the covalent coupling between terminal alkynes on a noble metal. J Phys Chem C 118:3181–3187

3. Liu J, Ruffieux P, Feng X, Müllen K, Fasel R (2014) Cyclotrimerization of arylalkynes on Au(111). Chem Commun 50:11200–11203
4. Zhou H, Liu J, Du S, Zhang L, Li G, Zhang Y, Tang BZ, Gao H-J (2014) Direct visualization of surface-assisted two-dimensional diyne polycyclotrimerization. J Am Chem Soc 136:5567–5570
5. Björk J (2016) Reaction mechanisms for on-surface synthesis of covalent nanostructures. J Phys Condens Matter 28:083002
6. Qi Y, Zhou X, Liu M, Li Q, Ma D, Zhang Y, Liu Z (2015) Controllable synthesis of graphene using novel aromatic 1,3,5-triethynylbenzene molecules on Rh(111). RSC Adv 5:76620–76625
7. Zhang Y-Q, Björk J, Weber P, Hellwig R, Diller K, Papageorgiou AC, Oh SC, Fischer S, Allegretti F, Klyatskaya S, Ruben M, Barth JV, Klappenberger F (2015) Unusual deprotonated alkynyl hydrogen bonding in metal-supported hydrocarbon assembly. J Phys Chem C 119:9669–9679
8. Eichhorn J, Heckl WM, Lackinger M (2013) On-surface polymerization of 1,4-diethynylbenzene on Cu(111). Chem Commun 49:2900–2902
9. Liu J, Chen Q, Xiao L, Shang J, Zhou X, Zhang Y, Wang Y, Shao X, Li J, Chen W, Xu GQ, Tang H, Zhao D, Wu K (2015) Lattice-directed formation of covalent and organometallic molecular wires by terminal alkynes on Ag surfaces. ACS Nano 9:6305–6314
10. Adisoejoso J, Lin T, Shang XS, Shi KJ, Gupta A, Liu PN, Lin N (2014) A singlemolecule-level mechanistic study of Pd-catalyzed and Cu-catalyzed homocoupling of aryl bromide on an Au(111) surface. Chem-Eur J 20:4111–4116
11. Zhao W, Dong L, Huang C, Win ZM, Lin N (2016) Cu-and Pd-catalyzed Ullmann reaction on a hexagonal boron nitride layer. Chem Commun 52:13225–13228
12. Varela JA, Saá C (2003) Construction of pyridine rings by metal-mediated [2+2+2] cycloaddition. Chem Rev 103:3787–3802
13. Abbet S, Sanchez A, Heiz U, Schneider W-D, Ferrari A, Pacchioni G, Rösch N (2000) Acetylene cyclotrimerization on supported size-selected pd n clusters ($1 \leq n \leq 30$): one atom is enough! J Am Chem Soc 122:3453–3457
14. Sorace L, Gatteschi D (2015) Electronic structure and magnetic properties of lanthanide molecular complexes. Wiley-VCH Verlag GmbH & Co, KGaA
15. Evans W, Allen N (2005) Divalent lanthanide reduction catalysts, US Patent 6,887,824
16. Kobayashi S (2003) Lanthanides: chemistry and use in organic synthesis. Springer, Berlin
17. Écija D, Auwärter W, Vijayaraghavan S, Seufert K, Bischoff F, Tashiro K, Barth JV (2011) Assembly and manipulation of rotatable cerium porphyrinato sandwich complexes on a surface Angew Chem Int Ed 50:3872–3877
18. Écija D, Urgel JI, Papageorgiou AC, Joshi S, Auwärter W, Seitsonen AP, Klyatskaya S, Ruben M, Fischer S, Vijayaraghavan S, Reichert J, Barth JV (2013) Fivevertex archimedean surface tessellation by lanthanide-directed molecular self-assembly.Proc Natl Acad Sci USA 110:6678–6681
19. Urgel JI, Cirera B, Wang Y, Auwärter W, Otero R, Gallego JM, Alcamí M, Klyatskaya S, Ruben M, Martín F, Miranda R, Ecija D, Barth JV (2015) Surface-supported robust 2D lanthanide-carboxylate coordination networks. Small 11:6358–6364
20. Rinehart JD, Long JR (2011) Exploiting single-ion anisotropy in the design of f-element single-molecule magnets. Chem Sci 2:2078–2085
21. Dreiser J (2015) Molecular lanthanide single-ion magnets: from bulk to submonolayers. J Phys Condens Matter 27:183203
22. Skomski R, Sellmyer D (2009) Anisotropy of rare-earth magnets. J Rare Earths 27:675–679
23. Donati F, Rusponi S, Stepanow S, Wäckerlin C, Singha A, Persichetti L, Baltic R, Diller K, Patthey F, Fernandes E, Dreiser J, Šljivančanin Ž, Kummer K, Nistor C, Gambardella P, Brune H (2016) Magnetic remanence in single atoms. Science 352:318–321
24. Feng J, Zhang H (2013) Hybrid materials based on lanthanide organic complexes: a review. Chem Soc Rev 42:387–410
25. Molander GA (1992) Application of lanthanide reagents in organic synthesis. Chem Rev 92:29–68

26. Nishiura M, Hou Z, Wakatsuki Y, Yamaki T, Miyamoto T (2003) Novel z-selective head-to-head dimerization of terminal alkynes catalyzed by lanthanide half-metallocene complexes. J Am Chem Soc 125:1184–1185
27. Zhang Y-Q, Björk J, Barth JV, Klappenberger F (2016) Intermolecular hybridization creating nanopore orbital in a supramolecular hydrocarbon sheet. Nano Lett 16:4274–4281
28. HyperChem^{TM} Release 7.0 (MolecularModeling), Hypercube, Inc. http://www.hyper.com/
29. Patterson CH, Lambert RM (1988) Molecular pathways in the cyclotrimerization of ethyne on palladium: role of the C4 intermediate. J Am Chem Soc 110:6871–6877
30. Hoffmann H, Zaera F, Ormerod R, Lambert R, Yao J, Saldin D, Wang L, Bennett D, Tysoe W (1992) A near-edge X-ray absorption fine structure and photoelectron spectroscopic study of the structure of acetylene on Pd(111) at low temperature. Surf Sci 268:1–10
31. Lomas JR, Baddeley CJ, Tikhov MS, Lambert RM (1995) Ethyne cyclization to benzene over Cu(110). Langmuir 11:3048–3053
32. Öberg H, Nestsiarenka Y, Matsuda A, Gladh J, Hansson T, Pettersson LGM, Östrm H (2012) Adsorption and cyclotrimerization kinetics of C2H2 at a Cu(110) surface. J Phys Chem C 116:9550–9560
33. ChemDraw Professional 15.0. CambridgeSoft. http://www.cambridgesoft.com/Ensemble_for_Chemistry
34. Kittelmann M, Rahe P, Nimmrich M, Hauke CM, Gourdon A, Kühnle A (2011) On-surface covalent linking of organic building blocks on a bulk insulator. ACS Nano 5:8420–8425

Chapter 7
Conclusion and Outlook

The focus of this thesis is the on-surface synthesis of alkyne-based nanostructures and characterization of their structural and electronic properties. The control and understanding of alkyne reactions toward organometallic and covalent structures on clean surfaces under well-defined laboratory conditions is an important step toward the synthesis of new interfaces with exquisite structure-function properties. The STM analysis of various molecular systems helped us to expand our knowledge on alkyne-based on-surface chemistry. In detail, different custom-designed molecules equipped with sp-hybridized functional groups, i.e., the protected alkynes ($C{\equiv}C{-}SiC_3H_9$) of HEB, terminal alkynes ($C{\equiv}C{-}H$) of Ext-TEB, and both terminal alkynes and $C{\equiv}N$ side groups of CN-DETP were studied on well-defined single-crystal silver substrates to initiate the on-surface formation of organometallic compounds and chains as well as covalent polymer structures.

Within the first two projects, we employed a novel precursor design (CN-DETP) with secondary functional groups besides terminal alkyne groups, in order to demonstrate the creation of sophisticated nanoarchitectures via heat-induced on-surface reactions. Firstly, the Ag(110) surface steers terminal alkyne reactions that lead to σ-like organometallic bonding motifs along cyano-functionalized Ag-bis-acetylide chains. CN-related side-chain recognition from secondary functional groups allows the periodic and densely packed stacking of organometallic chains, which results in enantiomeric 2D domains of trans isomers. Complementary XS spectroscopy confirms the presence of organometallic C(sp)–Ag–C(sp) bonding causing a significant hybridization between the alkynes' in-plane π^* orbitals and coordinated Ag atoms, and indicates the prevailing alignment of 1D chains, consistent with the STM results. The densely packed Ag-coordinated structures could express interesting optoelectronic properties relevant for light-emitting interfaces and photovoltaic devices.

The on-surface reaction products not only depend on the employed precursor design, but also on the structural and chemical properties of the surface. When studying the same molecule CN-DETP on the Ag(111) surface instead of Ag(110), we observed on-surface coupling via butadiyne bridges instead of organometallic

R. Hellwig, *Alkyne-Based Nanostructures on Silver Substrates*, Springer Theses,
https://doi.org/10.1007/978-3-030-00997-7_7

linkage. The chemoselectivity toward the targeted graphdiyne nanowires is significantly higher compared to the DETP precursor without secondary functional groups. Accordingly, we presented an approach toward the controlled creation of 1D covalent structures exhibiting CN side groups, whose interchain interactions lead to room temperature-stable double-strand assemblies. After the controlled isolation of extended graphdiyne nanowires via STM tip-mediated manipulations, we demonstrated their bending flexibility and the controlled cis $\leftrightarrow$ trans switching of individual CN-phenyl groups. The only byproducts limiting the length of the graphdiyne wires are Γ-shaped vinyl-ethynyl motifs, which were further reduced by employing the Ag(100) surface (instead of Ag(111)) for the on-surface coupling reaction. The high conformational flexibility and switching characteristics of CN-functionalized GDY nanowires bears high potential for applications in molecular electronics.

In addition to investigations on terminal alkyne reactions, a molecule with protected alkyne groups was investigated. The de novo synthesized sixfold-symmetric precursor HEB represents multiple binding sites for organometallic bonding, while the capping groups inhibit on-surface coupling reactions between monomers. By the co-deposition of HEB monomers and metal atoms, i.e., the 3d block element Co and 4f-block element Ho, we created polynuclear metal alkynyl π complexes on Ag(111). The on-surface formed dinuclear species (C_2) hosts two Co atoms, each confined between two adjacent alkynes in a tweezer-like bonding geometry. Interestingly, the periodic superlattice of Co atoms embedded within domains of assembled complexes (Co-HEB) and HEB units expresses an Archimedian 3.4.6.4 surface tesselation. Extensive DFT-calculations substantiate our picture of a chiral complex forming enantiomeric domains through epitaxy-guided interfacial coupling. Furthermore, STM tip manipulations on the isolated C_2 species indicate enantiomeric switching between its two conformer states. In addition, the tip-induced activation of a Kondo resonance was demonstrated for Co-HEB, whose Co atoms represent multiple spin centers embedded within a custom-designed alkyne precursor.

For comparison, the 4f-block element Ho expressing a significantly larger ionic radius than Co was utilized for establishing metal alkynyl complexes. The resulting organolanthanide complex engages the alkynes' π orbitals in a bonding scenario, in which incorporated Ho atoms sit on top of the alkyne moieties of HEB, as compared to the Co atoms of Co-HEB lying below them. The on-surface formed Ho-HEB species C_1, C_2 and C_3 are rationalized by the site-selective attachment of one, two and three Ho atoms above the sp-unsaturated periphery of a HEB monomer. In contrast to the epitaxy-induced periodicity of Co atoms within organocobalt islands, the mixed 2D pattern comprised of assembled Ho-HEB complexes and HEB monomers does not exhibit a periodic superlattice of embedded Ho atoms.

Intrigued by the formation of polynuclear Ho alkynyl π complexes with protected alkynes, lanthanide-based on-surface reactions with terminal alkyne groups (higher reactivity) were explored. For the first time, we presented Ho-catalyzed on-surface coupling with terminal alkyne monomers. Ho dosage at substrate temperatures down to $T \sim 130$ K leads to Ho-induced dimerization between predeposited Ext-TEB monomers on Ag(111). For Ho dosage at slightly higher substrate temperatures ($\sim$200 K), oligomeric structures formed via 1,2,4- and 1,3,5-cyclotrimerization were identified. Accordingly, we proposed a Ho-activated reaction route, in which dimeric vinyl-ethynyl motifs represent the intermediates for sp^2-unsaturated cyclotrimerization products.

Within this thesis, different strategies for the on-surface fabrication of covalent polymer structures and metal alkynyl complexes on silver surfaces were presented. Future experiments should target the in-depth study of their electronic properties via STS and complementary XS techniques, which can open up new vistas in the field of carbon-based molecular electronics. Such measurements require an insulating layer, in order to reduce electron interactions with the metal surface. To this end, one could grow insulating islands (e.g., NaCl) on a silver substrate and use the STM tip to lift an isolated Co-HEB complex, Ag-coordinated chain or extended graphdiyne nanowire from the metal surface on top of a NaCl island. A further approach worth considering would be the investigation of alkyne reactions toward polymer structures via molecular depositon (and metal dosage) on top of an insulating layer (e.g., graphene or hexagonal boron nitride) covering the entire metal surface. Both strategies may allow the comparison of conductance measurements along the polymer backbone of CN-functionalized Ag-bis-acetylide and GDY-like strands. Furthermore, the bandgap characteristics of individual molecular nanowires could be measured as a function of polymer length and strain, e.g., for different bending radii of the molecular wire, and compared with theoretical predictions on 1D GDY.

The dinuclear Co-HEB and Ho-HEB complexes may represent attractive spin-orbit materials with promising luminescent and magnetic properties, which may be tuned by employing different metal elements and alkyne-based building blocks. Transport experiments, spin-resolved spectroscopy and complementary XS synchrotron techniques could deliver interesting insights on the coupling between multiple redox-active and paramagnetic metal centers within the polynuclear compounds, which might lead to new ideas for the development of chemical sensors, optoelectronic devices and magnetic single-molecule switches. Combining on-surface coupling and organometallic bond formation with 3d/4f metal atoms on Ag(111), one could construct covalent polymer architectures embedding metal atoms via site-selective alkyne-metal-alkyne bonding.

The catalytic behavior of Ho in activating on-surface coupling at low temperatures could be investigated for different 4f-block elements and on different substrates, in order to unravel the role of the surface and catalyst on the reaction pathway and regioselectivity (1,3,5- versus 1,2,4-cyclotrimerization). It would also be interesting to study the reaction yield for cyclotrimerization as a function of the Ho cluster size. Furthermore, a one-atom thick Ho film supported on an inert substrate may represent a well-defined catalytic template for further optimization and control of this low-temperature coupling reaction.

Printed by Printforce, the Netherlands